# Topics in Intelligent Engineering and Informatics

## Volume 12

**Series editors**

János Fodor, Budapest, Hungary
Imre J. Rudas, Budapest, Hungary

More information about this series at http://www.springer.com/series/10188

László Nádai · József Padányi
Editors

# Critical Infrastructure Protection Research

Results of the First Critical Infrastructure
Protection Research Project in Hungary

 Springer

*Editors*
László Nádai
Óbuda University
Budapest
Hungary

József Padányi
National University of Public Service
Budapest
Hungary

Topics in Intelligent Engineering and Informatics
ISBN 978-3-319-80269-5      ISBN 978-3-319-28091-2   (eBook)
DOI 10.1007/978-3-319-28091-2

Printed on acid-free paper

This Springer imprint is published by SpringerNature
The registered company is Springer International Publishing AG Switzerland

# Foreword

The unity and complementary roles of education and research in the life of a higher education institute should not be questioned. Where there is research, there are new achievements. These achievements can be shared, in the education process with our students. This is not a new revelation; the mission of universities, from their foundation, is to transmit new achievements and knowledge, collect new results, and to conduct independent research.

Research cooperation—the summary of which the reader is holding in their hands—between Obuda University and Zrínyi Miklós National Defense University (the legal successor of the National University of Public Service) was established following these principles. Cooperation between the two universities did not start with this research, since we have a decade of common work together. It started with an educational cooperation in the field of security technology and then continued, with the common development of a Ph.D. curriculum. It was brought to its highest level within the framework of the research concerning the defense of critical infrastructures.

In the initial phase of setting up the research program, it was discovered that disciplines, scientific fields, and research areas, common to the two universities, showed a significant overlap in many areas. The previously mentioned security technology and the development and application of autonomous airborne equipment (no onboard pilot), generated challenges that resulted in institutional climate change, customized metal fabrication methods and new behavioral patterns for emergency situations. These are a few examples of the common fields of special interest for both of the universities and include research areas concerning technical sciences, security sciences, military technical sciences, and martial sciences that have strong overlap, and complementary and mutual enhancement possibilities.

Shared research topics have also reinforced the certainty that our teachers, researchers, and students (at least during common work) speak a common language, which is continuously progressed, by the recognition and pleasure of obtaining new relevant knowledge. We have discovered that solving the problems and above all, answering common questions brings great satisfaction and teamwork to the normally unrelated colleagues, such as engineers, psychologists, and soldiers.

We consider it a great achievement and important indicator that our B.Sc. and Ph.D. students have also participated in the research. For many of them, this was the first step toward a doctorate degree and/or a scientific career path within the university setting.

In our research area, we have efficiently and fully reached our goals. All undertakings and assumptions were fulfilled, and we are very proud of the results. Dissemination activities were especially prolific. Dozens of articles and books have been published to communicate our results, and we have presented the various findings at many conferences.

The specific project implementation was not lacking in disputes. We had many debates over personal as well as professional matters. It is not even sure that we always came to the right decision—today we can see what we should have done, perhaps in another way. But applying the guidelines of professional honor has always helped us in taking the next step and moving from an impasse. The amendment procedures that were a consequence of the termination of the Zrínyi Miklós National Defense University and succeeded by the National University of Public Service have not made our work easy. The transformation of administrative bodies has led to several changes that have affected our research work.

In seeing the achievements of the project, we cannot have another goal, but to continue on our current track. We have to identify the research fields where we can carry on with the work and also those fields where new opportunities are available. Knowing the commitment of our teachers and seeing the efforts of both universities, we are quite sure that we have the potential to maintain and reach the premier research results.

# Contents

# The First Critical Infrastructure Protection Research Project in Hungary

Tibor Babos

*Learning is not compulsory, neither is survival.*

W. Edwards Deming

**Abstract** Óbuda University and the National University of Public Service, applied together, for the Hungarian Government tender titled "*TÁMOP-4.2.1.B-11/2/KMR-2011-0001 promotion of research projects and research services in Central Hungary*" in 2011. As a result of the positive tender evaluation they won altogether ~3,167,000 € financial support from which the ÓU could use ~1,749,864 € and the NUPS could use ~1,417,143 € to accomplish the approved research objectives. After the successful tender the two beneficiaries created a consortium named "*Critical Infrastructure Protection Researches*". With the realization of the Project the employment of 112 lecturers, 30 experts, 28 future lecturers, 33 foreign experts, the publishing of 132 articles, 11 books, the submission of 4 patents, the organization of 140 conference-presentations, the development of 43 studies and strategies and the writing of 70 other scientific dissertations were set as a goal. The above requirements have been accomplished in some areas significantly over fulfilled during the 27-month course of the Project. The results of the research have been received with sincere renown in professional, academic, governmental and social circles.

**Keywords** Critical infrastructure protection · European Union research project · Hungarian scientific results · National security · Új Széchenyi Plan · Óbuda University · National University of Public Service

W. Edwards Deming Quotes, URL: http://www.brainyquote.com/quotes/authors/w/w_edwards_deming.html#Vp4MldA51AqCIisT.99 (Access Date: 10 April 2014).

T. Babos (✉)
Óbuda University, Budapest, Hungary
e-mail: babos@uni-obuda.hu

© Springer International Publishing Switzerland 2016
L. Nádai and J. Padányi (eds.), *Critical Infrastructure Protection Research*,
Topics in Intelligent Engineering and Informatics 12,
DOI 10.1007/978-3-319-28091-2_1

1

# 1 Introduction

In the everyday life of modern society, technical structures, especially in the area of information technology system maintenance, controlling and operating the energy-supply, drinking water resources and transportation, bear definitive importance. The evolution accelerated as a result of the globalization and the revolution of information technology turned these systems complex, interdependent, and at the same time, especially fragile and vulnerable. The political, economic, social and environmental events of the 21st century, particularly the terrorist attacks at New York and Washington on 11 September 2001; the terrorist actions in Madrid on 11 March 2004; the global economic recession that started in 2007; the devastating earthquakes in Turkey on 17 August 1999, then later on 10 June 2012 or in Haiti on 12 January 2010; the nuclear catastrophe at Fukushima on 11 March 2011; or the fires, as a result of the 2010 and 2013 heat waves in Russia, all of them made humanity realize that how defenseless the artificially created civilized world is to theses various disasters, either caused by human beings or nature. Nevertheless, these events highlighted the interdependence of infrastructures, society and the governmental function.

The Deming quote in the motto says it undeniably: survival depends on the ability of knowledge-based adaptation. It is essential from the aspect of the survival, safety and the prosperity of a nation or a state that how its critical systems are organized, managed and protected. In other words: the components of critical infrastructure have to withstand any kind of effect or change shall it be political, economic, social or environmental in order to secure the basic life and management conditions of the nation, the state and the government. Nevertheless, the principles, plans and requirements regarding critical infrastructures are not only important for the national administration of a particular nation, these have to be compatible with federal strategic interests and in Europe with the strategic interests of the EU.

Óbuda University (ÓU) and the predecessor of the National University of Public Service, the Zrínyi Miklós National Defense University (ZMNU) in 2011, commenced assembly of a wide-scale and innovative research project with the common goal-system of researching and studying the general and specific interrelations of critical infrastructures in an increasingly complex and vulnerable political environment and supplying the national and governmental decision makers with results that are scientifically established as well.[1] In the light of the foregoing, the objective of this study is to present the Project Nr. TÁMOP-4.2.1.B-11/2/KMR-2011-0001 titled *"Critical infrastructure protection research"* conducted between the 1st January 2012 and 31st March 2014 in a consortium cooperation of the ÓU and the NUPS. In the interest of reaching the intended aim, this article examines the

---

[1]Critical infrastructure protection researches, Critical infrastructure protection researches "TÁMOP-4.2.1.B-11/2/KMR-2011-0001" project, Óbuda University, Budapest, URL: http://news.uni-obuda.hu/articles/2014/03/20/kritikus-infrastruktura-vedelmi-kutatasok (Access Date: 28 April 2014).

foundations, circumstances, professional and financial features, structures and the results of the Project as well. The study—for the sake of understanding the topic—also discusses the basic interrelations of critical infrastructures and the international defense political environment in necessary detail, and takes a look at the trends and possible improvements of the topic.

## 2 Aftermath and Reaction of 9/11

On September 11, 2001, 19 militants connected to the Islamic radical group al-Qaeda hijacked four airliners and carried out suicide attacks against targets in the United States. Two of the planes were flown into the towers of the World Trade Center in New York City, a third plane hit the Pentagon just outside Washington, D.C., and the fourth plane crashed in a field in Pennsylvania. Often referred to as 9/11, the attacks caused extensive death and destruction, triggering major U.S. initiatives to combat terrorism and defining the presidency of George W. Bush. Over 3000 individuals were murdered during the attacks in New York City and Washington, D.C., including more than 400 police officers and firefighters.[2] Laymen associate the emergence of the issue with 11 September 2001, as this was the point when the public was informed by the media concerning the kind of harm that a carefully prepared series of terrorist attacks can inflict at the function of critical infrastructures, and how it paralyzes the governmental machinery and the everyday of civil life—even—in the most powerful of nations, at that time.[3]

It is a fact that the National Security Strategies created by numerous states in Europe are connecting the importance of the protection of critical infrastructures or the start of the modern timeline of the whole topic to the attacks against New York and Washington.[4] The assaults against the Twin Towers and the Pentagon—as it was experienced by all of us—reshaped everyday life of the developed world. Today we travel differently by airplane, we enter in another way into foreign countries and we bank in a different way. The governmental and national administration systems are built up according to a new logic and function with new mechanisms. As a result of September 11 the defense services, especially the secret

---

[2]9/11 Attacks, History.com Staff, 2010, A+E Networks, URL: http://www.history.com/topics/9-11-attacks (Access Date: May 10, 2014).

[3]Jim Garamone, Panetta Spells Out DOD Roles in Cyber defense, American Forces Press Service, Washington, October 11, 2012, URL: http://www.defense.gov/news/newsarticle.aspx?id=118187 (Access Date: 10 May 2014).

[4]Ten Years After 9/11, An Overview of New York State's Homeland Security Accomplishments, The New York State Decision of Homeland Security and Emergency Services, URL: http://www.dhses.ny.gov/media/documents/ten-years-after-9-11-nys-accomplishments.pdf (Access Date: 11 May 2014).

services, became significantly strengthened, old ministry subsystems vanished and new ones came to their place.[5]

After the terrorist attacks the transformation of the governmental structure was immediately started in the United States. A couple of months after the events radically new governmental mechanisms, structures and procedures came into effect. The exact operation of this system, which is much tighter, firmer and more integrated than the previously, is classified and therefore it is unknown for the public opinion and the researchers. The determining factor of the governmental structure transformation that is available from open sources is the integration of the activities of the highest governmental and security agencies. Today the top-secret world the government created in response to the terrorist attacks of 9/11, has become so large, so unwieldy and so secretive that no one knows how much money it costs, how many people it employs, how many programs exist within it or exactly how many agencies do the same work. After more than a decade of unprecedented spending and growth, the result is that the system put in place to keep the United States safe is so massive, that its effectiveness is impossible to determine.[6]

The Department of Homeland Security (DHS) founded in 2003 today lies above all armed and secret agencies that are related to the internal safety of the United States or the protection of its citizens. Of course this governmental institution is not the same as the Department of Defense as the latter is responsible for providing the military forces needed to deter war and protect the security of the country.[7] Accordingly, the physical safety of the nation and the armed protection of the country is still the responsibility of the Pentagon and the subordinated military forces. The primary responsibility of the new ministry is the protection of the United States and the territory of it from non-military aggression that is for example from terrorist attacks, industrial accidents and natural disasters. In case of the DHS the definition of protection includes preparation for all the possible hazard-sources, prevention, managing the occurred events, coordinating the government and after-caring the consequences.[8]

Not even one and a half years after the 2001 terrorist attacks, on 1st March 2003 the government authorities responsible for citizenship and naturalization; the entire customs, including the armed custom protection; the animal and plant health

---

[5]Billy Kenber, Outgoing Director Robert S. Mueller III tells how 9/11 reshaped FBI mission, National Security, The Washington Post, August 23, 2013, URL: http://www.washingtonpost.com/world/national-security/outgoing-director-robert-s-mueller-iii-tells-how-911-reshaped-fbi-mission/2013/08/22/ee452170-0b54-11e3-9941-6711ed662e71_story.html (Access Date: 10 May 2014).

[6]Diana Priest, William M. Arkin, A hidden world, growing beyond control, The Washington Post, July 19, 2010, URL: http://projects.washingtonpost.com/top-secret-america/articles/a-hidden-world-growing-beyond-control/ (Access Date: 11 May 2014).

[7]Mission Statement, Department of Defense, 1994, URL: http://govinfo.library.unt.edu/npr/library/status/mission/mdod.htm (Access Date: 10 May 2014).

[8]The National Strategy for The Physical Protection of Critical Infrastructures and Key Assets, the White House, Washington, USA, URL: http://www.dhs.gov/xlibrary/assets/Physical_Strategy.pdf (Access Date: 29 April 2014).

service; border control as a whole, together with the land, maritime border security and the reconnaissance; the decision preparation, and law enforcement armed forces supporting the function of the bodies above, all of them, merged into the DHS. The DHS is responsible for energy security and the critical infrastructure protection of the USA as well. Its responsibilities and competencies cover the entire territory of the United States, including areas or zones of interest falling under its jurisdiction or auspices.[9] Accordingly, its offices and staff can be found in all U.S. embassies, in all allied countries.

On December 17th, 2003, President Bush signed Directive HSPD-7, "Critical Infrastructure Identification, Prioritization and Protection". This direction supersedes Presidential Decision Directive/NSC-63 of May 22, 1998, "Critical Infrastructure Protection", and establishes a national policy for Federal departments and agencies to identify and prioritize United States critical infrastructure and key resources and to protect them from terrorist attacks. The Directive instructed federal departments and agencies to prepare plans for protecting physical and cyber critical infrastructure and key resources, owned or operated, including leased facilities by July 31, 2004. This project, developed in consultation with the Homeland Security Council and the DHS, includes the required format for agencies to use when submitting internal critical infrastructure protection plans. Pursuant to the guidance these plans had to address identification, prioritization, protection, and contingency planning, including the recovery and reconstitution of essential capabilities. In particular, planning must include protection priorities, the agency's ability to ensure continuity of business operations during a physical or cyber attack, and, where current capabilities are lacking, plans of action and milestones to achieve the necessary level of performance. Upon submission, security plans got to the subject to an interagency review coordinated by DHS. The goals of these reviews included ensuring consistent planning and protection of Federal critical infrastructure and key resources across the Federal government.[10] This memorandum determined the leading role of the DHS in all security related issues in the highest American decision making.

Today with its nearly 200,000 employees and a budget of 56.3 Billion dollars it is the United States' third largest and one of its most influential departments. It gives rise to interesting conclusions, that today the allotted budget of the DHS is more than 60 billion dollars. Therefore regarding its tasks and governmental

---

[9]Written testimony of CBP Deputy Commissioner Kevin McAleenan and ICE Deputy Director Daniel Ragsdale for a House Committee on Homeland Security, Subcommittee on Border and Maritime Security hearing titled "Authorizing Customs and Border Protection and Immigration and Customs Enforcement", Review Date: April 7, 2014, Department of Homeland Security, URL: http://www.dhs.gov/news/2014/04/08/written-testimony-cbp-and-ice-house-homeland-security-subcommittee-border-and (Access Date: 10 May 2014).

[10]Development of Homeland Security Presidential Directive—7 Critical Infrastructure Protection Plans to Protect Federal Critical Infrastructures and Key Resources, M-04-15, Memorandum for the Heads of Executive Departments and Agencies, Executive office of the President, Office of Management and Budget, Washington D.C., June 17, 2004, URL: http://www.whitehouse.gov/sites/default/files/omb/memoranda/fy04/m-04-15.pdf (Access Date: 11 May 2014).

importance the DHS shows numerous similarities to the nature of the European ministries of internal affairs, however in some aspects it might even exceed them. The complete transformation of the central national administration escalated further onto federal, state, county, regional and even local governmental levels.


## 3 The "Dawn" of Critical Infrastructure Protection

It has to be stated that "the protection of critical infrastructure" should not be connected to 11 September 2001, it dates back even much further in time. The available sources in written history clearly show that every country and nation meticulously protected their own week points, they took care of the protection of their critical infrastructure, although the definition had not existed in this form. The oriental, Egyptian, Greek or Roman empires alike, strictly protected their main transportation networks, food-supply routes, material resources or management techniques and kept them in the greatest secret. Quality improvement in the "protection of critical infrastructure" took place in 1648 after the Peace Treaties of Westphalia with the establishment of nation-states, when states got the necessary governmental tools and structures at their own disposal, and as state systems; structures and mechanisms similar to ours in nowadays were created.[11] They handled their critical systems in a similar sense as today and defended them with means to our recent "modern" tools. Today's "protection of critical infrastructure" is the "product" of the two world wars and the Cold War.

The term "critical infrastructure protection" has been first used by U.S. president Bill Clinton in his presidential directive issued on 22 May 1998. In its introduction he observed that potential vulnerability of the United States is increasing, therefore, the protection of critical infrastructure deserves increased attention. He noted that the Unites States possesses both the world's strongest military and its largest national economy.[12] These two aspects of power are mutually reinforcing each other but they are dependent on each other as well. These aspects are increasingly reliant upon critical infrastructures and information technology systems. Critical infrastructures are those physical and virtual systems that are essential to the minimum operation and function of the social life, economy and the government. These include but are not limited to the following areas: telecommunications; energy; banking and finance; transportation; water systems (Table 1).

According to the text the President made a commitment that the United States shall take all necessary measures to ensure the uninterrupted operation and protection of the physical and virtual critical infrastructures, and it shall prosecute any

---

[11]Tibor Babos, The Five Central Pillars of European Security, NATO Public Diplomacy Division, Brussels, 2007, p. 14.

[12]Critical Infrastructure Protection, Presidential Decision Directive NSC-63, The White House, Washington, May 22, 1998, URL: http://www.fas.org/irp/offdocs/pdd/pdd-63.htm (Access Date: 11 May 2014).

**Table 1** The first and last page of the critical infrastructure protection, presidential decision directive NSC-63 (edited by: Babos Tibor)

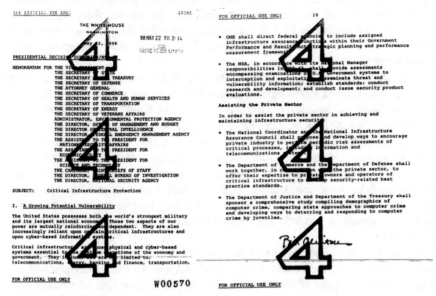

Critical Infrastructure Protection, Presidential Decision Directive NSC-63, The White House, Washington, May 22, 1998, URL: https://www.fas.org/irp/offdocs/pdd/pdd-63.pdf (Access Date: 11 May 2014)

action that could be a cause of vulnerability of these structures. The third section of the Presidential Decision Directive NSC-63 on Critical Infrastructure Protection states that not later than the year 2000, the United States shall have achieved an initial operating capability and no later than five years from today the United States shall have achieved and shall maintain the ability to protect the nation's critical infrastructures from intentional acts that would significantly diminish the abilities of:

- The Federal Government to perform essential national security missions and to ensure the general public health and safety
- State and local governments to maintain order and to deliver minimum essential public services
- The private sector to ensure the orderly functioning of the economy and the delivery of essential telecommunications, energy, financial and transportation services.[13]

The directive dedicates a whole separate subtitle to the support of private sector. It sets forth that in order to reach infrastructure security the competent bodies shall develop a framework of objectives that would encourage private companies to

---

[13]See Footnote 12.

conduct periodic inspections. It orders the public bodies to work together with private sector corporations in order to work out the best procedures and standards through professional debate. The financial and the justice departments were appointed to be in charge. In their care a comprehensive study had to be prepared which presented all the relevant challenges that could cause threat to critical infrastructures.

Retrospectively, the fourth section of the Directive might seem as a forecast or as a vision of a complex attack against critical infrastructures of the United States. It says that since the targets of attacks on the United Sates' critical infrastructure would likely include both facilities in the economy and those in the government, the elimination of the potential vulnerability requires a closely coordinated effort of both the government and the private sector. To succeed, this partnership must be genuine, mutual and cooperative. In seeking to meet the national goal to eliminate the vulnerabilities of the United States' critical infrastructure, therefore, the American Government should, to the extent feasible, seek to avoid outcomes that increase government regulation or expand unfunded government mandates to the private sector.[14]

For each of the major sectors of the American economy that are vulnerable to infrastructure attack, the Federal Government will appoint from a designated Lead Agency a senior officer of that agency as the Sector Liaison Official to work with the private sector. Sector Liaison Officials, after discussions and coordination with private sector entities of their infrastructure sector, will identify a private sector or counterpart (Sector Coordinator) to represent their sector. Together these two individuals and the departments and corporations they represent shall contribute to a regional National Infrastructure Assurance Plan by:

- Assessing the vulnerabilities of the sector to cyber or physical attacks
- Recommending a plan to eliminate significant vulnerabilities
- Proposing a system for identifying and preventing attempted major attacks
- Developing a plan for alerting, containing and rebuffing an attack in progress and then, in coordination with FEMA as appropriate, rapidly reconstituting minimum essential capabilities in the aftermath of an attack.[15]

It can be said without exaggeration that the Clinton-directive summed up the topic according to still valid criteria and definitions. A particular value of this basic document is the phrasing according to which "the elements of critical infrastructure include but are not limited to" the elements listed there. Since a result of technical development everyday new infrastructural elements will be listed which have not been there before. A further virtue of the directive is that it did not just invite the various non-public actors, market and financial potentates into governmental decision-making but made them interested in joining the system that was to be established by the government. The leadership of the United States has therefore

---

[14]See Footnote 12.
[15]See Footnote 12.

about one and a half decade ago noted: it does not matter if he rules the world economy and military wise, technological development made him so vulnerable, first of all in the areas of communication, transportation, resource supply, energy and other networks, that as a result of intentional attacks or other events of loss even the whole public administration and everyday life could become totally paralyzed. This was the reason why the USA handled the terrorist attacks of 11 September 2001 against New York and Washington with relatively good performance, quickly, thoroughly and operatively. In other words, the presidential directive issued previously, about one and a half year before and the numerous measures taken as a result of it protected the people and the government of America with a "well established protocol" against terrorist attacks.

## 4  Critical Infrastructure Protection Development

The terrorist events of 11 September 2001 made a huge impact on the world's public opinion and caused significant changes in world politics. Nine days after the attack the Bush government announced the Global War on Terror strategy (GWOT), which was soon adopted by all the western democracies and by some other countries, for example by almost all the Eastern-European Countries, Pakistan or the United Arab Emirates. The GWOT became a defining tendency in international relationships.[16] This community dominated by the United States as a result of 11 September started an international political, economic and military campaign against al-Qaeda and other militant organizations. Public enemy number one was Osama bin Laden, the operations directly targeted the countries where radical Islamic groups resided, and indirectly affected all Muslim countries. The international campaign was approved by the UN, NATO became its leading international organization, however, the OSCE and the EU took significant separate measures as well.[17]

NATO has been regulating and strictly protecting its critical infrastructures since its foundation. According to the Funding Document of the Alliance the five potential scenarios in which NATO is supposed to play a role are the following:[18]

1. Supporting Alliance military operations under Article 5
2. Supporting non-Article 5 crisis response operations
3. Supporting national authorities in civil emergencies

---

[16]Sherwood Ross: Rendition and the "Global War on Terrorism": 28 Nations Have Supported the US in the Detention and Torture of "Suspects", Global Research, April 1, 2010, URL: http://www.globalresearch.ca/rendition-and-the-global-war-on-terrorism-28-nations-have-supported-the-us-in-the-detention-and-torture-of-suspects/18419 (Access Date: 27 April 2014).

[17]ISAF's Mission, NATO and Afghanistan, North Atlantic Treaty Organization, URL: http://www.globalresearch.ca/rendition-and-the-global-war-on-terrorism-28-nations-have-supported-the-us-in-the-detention-and-torture-of-suspects/18419 (Access Date: 27 April 2014).

[18]For an exhaustive description of NATO's role in emergency planning consult NATO Handbook, Public Diplomacy Division, Brussels, 2006, pp. 297–302.

4. Supporting national authorities in the protection of their populations against the effects of WMD
5. Co-operation with Partners in the field of Civil Emergency Planning

According to the protocol created in the Cold war NATO ensures the safety of the critical infrastructure elements of its allies, and its member states. In order to ensure a coordinated approach for Civil Emergency Planning (CEP), the key role was assigned to the Senior Civil Emergency Planning Committee (SCEPC), reporting directly to the North Atlantic Council (NAC). CEP is an important activity in the foresight of disaster relief and is aimed at coordinating national resources. In the context of natural and man-made disasters, agreements enshrine NATO's role in the emergency setting. As an example, the "NATO Policy on Disaster Assistance in Peace Time" of May 9, 1995 and the statement "Enhanced Practical Cooperation in the field of Disaster Relief" of May 29, 1998 can be mentioned. In addition, NATO's Strategic Concept of 1999 acknowledges major disaster as a source for security and stability concerns.[19]

The term "critical infrastructure protection"—according to the Clinton-directive mentioned above—after the terrorist attacks of 11 September 2001 was immediately entered onto the schedule of the North Atlantic Council. Though as a result of the attacks against New York and Washington the 5th Article was not imposed for the expressed request of the USA, the Organizations reaction was manifested both politically and military wise. In the aftermath of the attacks of September 11, 2001, the NATO Prague summit initiated the "Civil Emergency Action Plan": a list of all available national resources was proposed, drawing the framework for assistance. In addition, exercises are planned to test and possibly improve interoperability. At the same time, the "Partnership Action Plan against Terrorism" was released.[20]

After 9/11, the preparedness of the Member States in areas of CIP (planning and infrastructure listings) was examined. The result was a Concept Paper on CIP, prepared by SCEPC. Key objectives are summarized in the exchange of information between stakeholders, assistance and development of training and education programs contributing to the identification of CI, determining research to support CIP and assistance during exercises. The "Planning Boards and Committees" (PB&Cs) of the SCEPC have started the necessary studies: national experts from government and industry, as well as military representatives are coordinating planning in eight technical domains: civil air transport, civil protection, food safety, industrial production and logistics, domestic surface transportation, medical affairs, shipping and finally civil electronic communications.[21] In April 2005 SCEPC adopted an adapted action plan in order to cover the efforts during and after CBRN terrorist attacks. The plan focuses on the protection of critical infrastructure (CI) and assistance to victims.

---

[19]Bart Smedts, NATO's Critical Infrastructure Protection and Cyber Defense, Royal High Institute for Defense Center for Security and Defense Studies, Focus Paper 19, July 2010, Brussels, p. 11.
[20]See Footnote 19.
[21]See Footnote 19.

In the midst of this, especially with increasing the activity of the European allies the terrorist attack in Madrid in March 2004, the cyber-attack against Estonia in 2007, the Russian-Georgian conflict in 2008, the pirate attacks continuous since 2008 in the Gulf of Aden and at the coasts of Somalia, and most recently the continuously escalating Russian-Ukrainian conflict played a significant role. NATO-operations in the last decade clearly show that the organization takes responsibility, has operation plans and if necessary intervenes in cases when NATO-interests—let it be ship routes, air corridors, pipelines, telecommunications or communication networks—are attacked outside of NATO-area, even in strategic distance from it.[22] Today it can be said, that NATO is not just in its conceptual and strategic documents about the protection of critical instruments, but it has policies and practices on operational level as well.

The European Council of June 2004 asked the Commission and the High Representative to prepare an overall strategy to strengthen the protection of critical infrastructure. As a complement to the measures that had been taken at national level, the EU has already adopted a number of legislative measures setting minimum standards for infrastructure protection in the framework of its different policies. This is notably the case in the transport, communication, energy, occupational health and safety, and public health sectors. A further step towards communication security was made with the creation of the European Network and Information Security Agency (ENISA). In addition, in sectors like aviation and maritime security, inspection services have been created within the Commission to monitor the implementation of security legislation by EU countries.[23]

In Europe the cooperation of the EU and NATO capabilities and procedures are a great opportunity for all national partners, both public services and private companies. Promising, that particular attention is given to certification of personnel and equipment, both for the protection of critical infrastructure in homeland security as well as for expeditionary forces. Military operations abroad are equally dependent on critical infrastructure, although other means of protection are used: preservation of supply and communication lines is essential to the success of operations for expeditionary forces.[24]

Hungary joins through its NATO membership, since 1999 and EU membership since 2004 directly to the international context of this topic. As a NATO and EU member, significant transformations took place in public administration here in our country as well as a result of the terrorist attacks and the following international campaign. It can be stated in general, that today, the Hungarian public administration operates in a new structure, in a more integrated way and according to more

---

[22]See Footnote 17.

[23]Critical infrastructure protection, Act, Communication from the Commission to the Council and the European Parliament of 20 October 2004—Critical Infrastructure Protection in the fight against terrorism [COM (2004) 702 final—Not published in the Official Journal], URL: http://europa.eu/legislation_summaries/justice_freedom_security/fight_against_terrorism/l33259_en.htm (Access Date: 28 April 2014).

[24]Bart Smedts, NATO's Critical Infrastructure Protection and Cyber Defense, Royal High Institute for Defense Center for Security and Defense Studies, Focus Paper 19, July 2010, Brussels, p. 27.

stringent measures. From the aspect of the security of the nation the departments responsible for the Hungarian Defense Forces, law enforcement bodies, bodies responsible for public administration and diplomacy corps and the Prime Ministry play the main role as the bodies of public administration, the armed forces, the secret agencies or other security institutions are subordinated to them. Considering that Hungary has not been affected by direct terrorist attacks, the Hungarian governmental decision making and implementations follow NATO and EU directives, with some delay.

# 5   Critical Infrastructure Protection Research Project Generation

The management of Óbuda University (ÓÚ) and the predecessor of the National University of Public Service, the Zrínyi Miklós National Defense University (ZMNU) realizing where the international trends of critical infrastructures were heading, and building on their own universities' capacity mutually initiated conceptual discussions in 2010 about a closer cooperation in this topic. As a result of the meetings they had come to the conclusion that relying on the technical and IT know-how of Óbuda University and the experience of the National University of Public Service in the fields of defense politics and military science they could jointly research the social and natural scientific aspects of the current problems of critical instruments.[25] Both from the perspective of system theory and cost effectiveness it was a priority task to secure the optimal and coordinated utilization of the available technologies for civil security, and furthermore to stimulate the cooperation between service providers and consumers of civil security solutions.[26] Therefore, the project was realized through the consortium cooperation of a military and a civil university lecturing and researching on the field, as the cooperation between the two universities looked back onto a decade-long history.

Assessing the national and international professional-academic needs and the currently available research tenders, they have formulated a concept in accordance with the 18 December 2006 Decision No. 1982/2006/EC of the European Parliament and Council to initiate a common scientific research project in the European Union.[27] The initial idea was that the two universities together would

---

[25]Analysis of the Field, Critical Infrastructure Protection Researches "TÁMOP-4.2.1. B-11/2/KMR-2011-0001" Project, Óbuda University, Budapest, URL: http://nik.uni-obuda.hu/s/ tamop/node/2 (Access Date: 28 April 2014).

[26]The Problem Justifying, the Necessity of the Project and the Strategic Direction, Critical Infrastructure Protection Researches "TÁMOP-4.2.1.B-11/2/KMR-2011-0001" Project, Óbuda University, Budapest, URL: http://nik.uni-obuda.hu/s/tamop/node/4 (Access Date: 28 April 2014).

[27]Analysis of the Supply and Demand Factors of the Development, Critical Infrastructure Protection Researches "TÁMOP-4.2.1.B-11/2/KMR-2011-0001" Project, Óbuda University, Budapest, URL: http://nik.uni-obuda.hu/s/tamop/node/3 (Access Date: 28 April 2014).

undertake to assemble, adapt and disseminate the internationally available knowledge, and would endeavor to create new knowledge and technologies that could assist securing the safety of the citizens of the European Union against threats such as terrorism, nature, artificial catastrophes or organized crime.[28]

The following milestone of the project can be dated onto February 2011 when Óbuda University and the Zrínyi Miklós National Defense University executed an agreement in order to deepen the academic and professional activities, to develop the relations, to further the theoretical and practical training in the institutions and to promote R+D+I+E activities, to keep up the direct information-exchange and to prepare common tenders. With the conceptual preparations and the collaboration agreement in hand in the spring of 2011 the two universities applied together for the tender titled *"TÁMOP-4.2.1.B-11/2/KMR-2011-0001 promotion of research projects and research services in Central Hungary."* As the positive result of the tender review the two universities won the realization of the project in the total sum of 989,531,511 HUF, from which Óbuda University could use 546,745,071 HUF and the National University of Public Service could use 442,786,440 HUF to accomplish the research goals of the Project.[29] Following the successful tender the two beneficiaries created a consortium named *"Critical infrastructure protection researches"*, and organized the management, professional, administration and financial systems of the Project.

The management of the two Universities set the general goal of the project to assemble, to adapt and to disseminate the internationally available knowledge and to create new conclusions, procedures, technologies, altogether knowledge, which would promote securing the safety of the citizens.[30] The professional-academic goals of the project on the field of critical infrastructure protection: The consolidation (1) and if necessary training (2) of a critical mass of human capacity necessary for research and development activities on international level and in international cooperation, and the support of innovation in this field (3). As a result the following tasks were set:

- The development of a domestic framework of research and a "term base", and making these accepted
- The analysis of the effects of the latest generation computation and communication technologies, and its integration into the governmental, professional and scientific strategies

---

[28]Project Description, Critical Infrastructure Protection Researches "TÁMOP-4.2.1. B-11/2/KMR-2011-0001" Project, Óbuda University, Budapest, URL: http://nik.uni-obuda.hu/s/ tamop/ (Access Date: 28 April 2014).

[29]Critical Infrastructure Protection Researches, Critical infrastructure protection researches "TÁMOP-4.2.1.B-11/2/KMR-2011-0001" Project, Óbuda University, Budapest, URL: http:// news.uni-obuda.hu/articles/2014/03/20/kritikus-infrastruktura-vedelmi-kutatasok (Access Date: 28 April 2014).

[30]See Footnote 27.

- Development of flexible, self-diagnosing, self-healing physical and cyber infrastructural elements (methods, processes, technologies, services)
- The insurance of the optimal and coordinated utilization of the existing technologies for civil and military safety, the support of the common civil and military scientific researches
- The stimulation of the cooperation between service providers and consumers of civil security solutions
- The rationalization and optimization of system theory and cost effectiveness processes

## 6 Project Structure and Mechanism

Structurally, the Project is divided into supervisory, management and executive bodies as it can be seen on Table 2.

The project was supervised by the Professional Advisory Board (PAB) consisting of the rectors, the competent vice-rectors and deans of the two universities. The PAB decided in every matter which considered the whole of the Project, or as a result of which the progress of the Project would deviate from its original schedule. The sessions were held in every semester, or were organized "ad hoc" by the management.

Consisting of seven members with a project manager at the lead the Management was responsible for the all-around leading of the Project. In its subordination the professional leader and the bursar managed the corresponding sub-areas, their work was assisted by a two-man secretariat with a project assistant and a financial assistant from both universities.[31] The Management held weekly sessions where every management member took part. The range of the participants at management-meetings was supplemented with the participation of the competent managers, experts and other guests of the two universities according to the agenda. The agendas of the meetings were organized according to the cases, issues and tasks necessary for the everyday management and operation of the Project.

The Management called together in every semester or when necessary a Project-session where the management members, subprogram leaders, the leaders of priority research areas and the university leaders and experts supporting the Project were invited. The project manager was the chairman at the management- and Project-sessions, minutes were made from the mentioned statements, tasks and results. These minutes were filed to the Professional Advisory Board in every case.

---

[31]Project Management, Critical Infrastructure Protection Researches "TÁMOP-4.2.1. B-11/2/KMR-2011-0001" Project, Óbuda University, Budapest, URL: http://nik.uni-obuda.hu/s/ tamop/menedzsment (Access Date: 28 April 2014).

**Table 2** The structure of the *"Critical Infrastructure Protection Research"* project (Tibor Babos)

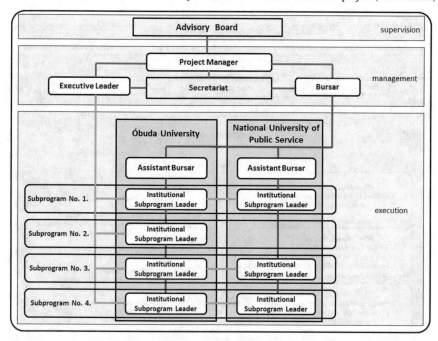

Project Structures, Critical Infrastructure Protection Researches "TÁMOP-4.2.1.
B-11/2/KMR-2011-0001" Project, Óbuda University, Budapest, URL: http://nik.uni-obuda.hu/s/
tamop/projekt-strukturak (Access Date: 28 April 2014)

Regarding the professional activity the Project focused on 21 priority research areas, which areas were organized—for the sake of a more integrated procession and a more operative Project-management—into four subprograms.[32] Each priority research area was managed by a leading professor and the programs were controlled by sub-program leaders. The priority research areas were the following:

1. Highly reliable, fault-tolerant, so-called "Self-healing" infrastructure sub-systems
2. Integrated management of data from the individual sub-systems
3. Distributed computing in sub-systems and secure communication between subsystems
4. Sustainable increase of the security level through the cooperation of state institutions, operators, owners and citizens.

The following Tables 3, 4, 5 and 6 illustrate the structure of the sub-programs.

---

[32]See Footnote 28.

**Table 3** The structure of sub-program no. 1. of the "*Critical Infrastructure Protection Research*" project (Tibor Babos)

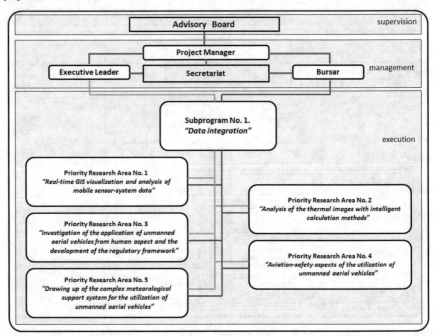

Project-structures, Critical infrastructure protection researches, TÁMOP-4.2.1. B-11/2/KMR-2011-0001 project, Óbuda University, Budapest, URL: http://nik.uni-obuda.hu/s/tamop/projekt-strukturak (Access Date: 28 April 2014)

## 7   Project Process

The period lasted as planned from 1 January 2012 until 31 December 2013. The supervising authority above the project was the European Social Fund (ESF) which made periodic and extraordinary on-site inspections on the Project verifying its compliance with the professional, financial and administrative rules; the proper development of the progress plans and the satisfaction of indicators. In accordance with the support agreement the supervising authority obligated the beneficiary to prepare professional reports in every 6 months. The management naturally implemented these obligations in four phases, in 6-month cycles and in accordance with the corresponding reports.

In the realization of the Project—besides the professional aspect—the economic/financial specialties' performance was determinative as without the planned, specialized, precise utilization of nearly 1 Billion HUF we could hardly talk about a successful Project. The Project filed accounting requests according to the deadlines predefined by the promoter periodically or as the progress yielded

**Table 4** The structure of sub-program no. 2. of the *"Critical Infrastructure Protection Research"* project (Tibor Babos)

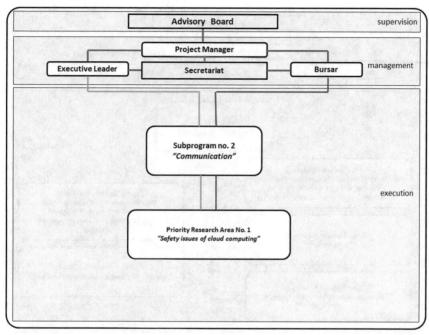

Project-structures, Critical infrastructure protection researches, TÁMOP-4.2.1. B-11/2/KMR-2011-0001 project, Óbuda University, Budapest, URL: http://nik.uni-obuda.hu/s/ tamop/projekt-strukturak (Access Date: 28 April 2014)

results. The Project was prepared for verification in every case and absolved these well, in most cases without need for rectification.

To ensure a high level of professional and financial closure of the Project the ESF approved a three-month extension for the physical implementation of the project, the maturity increased from 24 to 27 months and the date of closure changed from 31 December 2013 to 31 March 2014. Altogether the ESF approved the partial utilization of the reserves as well, and as a result the Project showed above 100 % performance from financial aspect too.

Basically, the following events can be considered milestones in the life of the Project:[33]

January 2012          Project-initiation
March 2012            PAB-meeting

---

[33]Critical Infrastructure Protection Researches, "TÁMOP-4.2.1.B-11/2/KMR-2011-0001" Project, Óbuda University, Budapest, URL: http://news.uni-obuda.hu/articles/2014/03/20/kritikus-infrast-ruktura-vedelmi-kutatasok (Access Date: 28 April 2014).

**Table 5** The structure of sub-program no. 3. of the *"Critical Infrastructure Protection Research"* project (Tibor Babos)

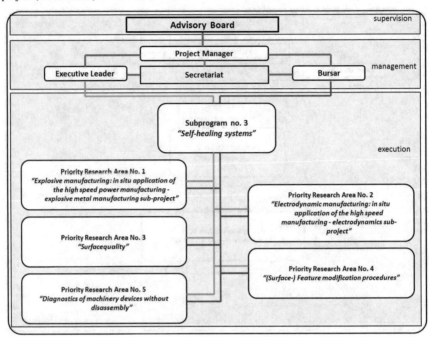

Project-structures, Critical infrastructure protection researches, TÁMOP-4.2.1. B-11/2/KMR-2011-0001 project, Óbuda University, Budapest, URL: http://nik.uni-obuda.hu/s/ tamop/projekt-strukturak (Access Date: 28 April 2014)

| | |
|---|---|
| May 2012 | Project opening event at the NUPS |
| July 2012 | General Report on the first semester |
| September 2012 | Annual International Conference with the assistance of the Ministry of Interior |
| January 2013 | General Report on the second semester; |
| July 2013 | General Report on the third semester; |
| 27 May 2013 | Public Forum |
| November 2013 | PAB-meeting |
| December 2013 | Request for the maturity extension and the utilization of reserves |
| January 2014 | General Report on the fourth semester |
| 19 March 2014 | Project Closing Event; |
| 31 March 2014 | Project-closure |
| From March 2014 | Project-accounting and availability |

**Table 6** The structure of sub-program no. 4. of the *"Critical Infrastructure Protection Research"* project (Tibor Babos)

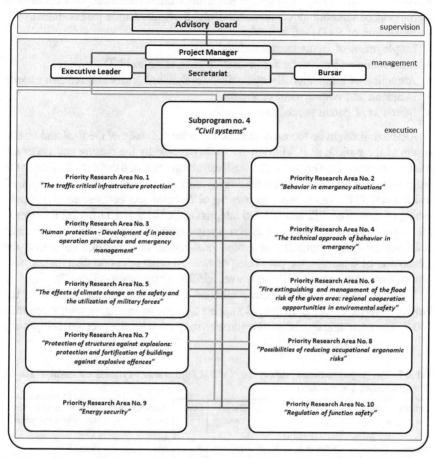

Project-structures, Critical infrastructure protection researches, TÁMOP-4.2.1. B-11/2/KMR-2011-0001 project, Óbuda University, Budapest, URL: http://nik.uni-obuda.hu/s/ tamop/projekt-strukturak (Access Date: 28 April 2014)

# 8  Project Indicators

The two Universities undertook to realize the following indicators:[34]

- Publishing articles in national and international papers
- Preparation and organization of conference-presentations

---

[34]Critical Infrastructure Protection Researches, TÁMOP-4.2.1.B-11/2/KMR-2011-0001 project, Articles of Incorporation, Óbuda University Budapest.

- Publishing national and international monographs
- Assistance in writing professional-academic studies and conceptual documents
- Support of scientific documents, Students' Scientific Circle papers, theses
- Expansion of R+D+I+E activities
- Employment of lecturers, researchers
- Mentorship and inclusion for students participating in Ph.D. training
- Attraction of internationally recognized foreign authorities and renown experts
- Inclusion and employment of experts
- Initiation of patent procedures

Although it might be too early to draw a precise summary of the final and factual results of the project, as it will only be made possible by the closing audit after the complete closure of the Project, however, it can already be stated that the indicators in the progress plan of the Project, that is the employment of 112 lecturers and researchers, 30 national and 33 foreign experts; the training of 28 future lecturer, researcher; the publishing of more than 132 national and international articles and 11 books; the organization and support of 140 national and foreign conference-presentations; the direct assistance in 43 governmental or professional-academic conceptual document; the submission of 4 patents; the publishing of over 70 other scientific dissertation, SSC papers, theses have been significantly over fulfilled by the project (Table 7).

The current data show that the professional and scientific performance of the Project exceeds even the envisaged support agreement indicators with more than 100 %. The economic specialty—in sync with the professional aspect—accomplished

**Table 7** The project indicator table of the "*Critical Infrastructure Protection Research*" Project (Tímea Scheller, Tibor Babos)

| Indicator subject | Target value | Unit | Accomplished value | Target value versus accomplished value (%) |
|---|---|---|---|---|
| Number of *articles* published in national and international professional *journals* with the assistance of the Project | 132 | Piece | 206 | 156.06 |
| Number of national and international *monographs* published with the assistance of the Project | 11 | Piece | 11 | 100.00 |
| Number of *R&D* activities realized with the assistance of the Project | 21 | Piece | 21 | 100.00 |
| Number of *lecturers* and *researchers* involved in the realization of the Project | 112 | Piece | 157 | 140.18 |
| Number of students enrolled in *doctoral* degree training participating in the Project | 28 | Piece | 35 | 125.00 |
| Number *patent* (domestic or international) *applications* (patent, protection) filed as the result of the Project | 4 | Piece | 4 | 100.00 |

Critical Infrastructure Protection Researches, "TÁMOP-4.2.1.B-11/2/KMR-2011-0001" Project, Óbuda University, Budapest, URL: http://news.uni-obuda.hu/articles/2014/03/20/kritikus-infrastruktura-vedelmi-kutatasok (Access Date: 28 April 2014)

excellently as well, as from one hand the utilization of costs carried were out in order, with discipline, and on the other hand with the use of the reserve besides the planned original ones new sources became accessible as well. This resulted in an above 100 % performance from financial aspect as well.

# 9  Summary

In conclusion, the scientific research Project lasting from 1 January 2012 through 31 March 2014, a period of 27 months, with a total cost of 989,531,511 HUF (~3,167,000 €) was concluded successfully from both professional and financial aspects and has reached its objectives. It has been clearly proven, with the conclusion of the Project, that the participating researchers and experts on the field of critical infrastructure protection have fulfilled the consolidation and development of a critical mass of human capacity necessary for research and development activities on international level and in international cooperation, and the support of innovation in this field. In accordance with the objectives and the monitoring indicators of the Project, Óbuda University and the National University of Public Service during the 27 months of the realization placed the emphasis on organization development, on human resource expansion, and on the publishing activity serving as a measure thereof.

Regarding the connections of the Project it can said that, the protection of critical infrastructure is not an isolated issue, the everyday function of modern societies and economies significantly depend on the reliable function and operation of infrastructure systems. Therefore, the Project joining to the objectives of the Új Széchenyi Plan contributed to the maintenance and increase of Hungary's and indirectly the EU's competitiveness, especially in relation to such "key sectors" as traffic informatics, geo informatics, logistics, info communication technologies.[35] Nevertheless, the critical infrastructure research is not just promoting the development of the education-research activity, the increase of educational standards or the growth of the ratio of certified lecturers at the two Universities. It is certain that the social utility of the research—from the aspect of creating the defense stability of Hungary and the EU—is really high, its effects will be felt long-term and by all levels of society.

*It is not enough to do your best; you must know what to do, and then do your best.*[36]

W. Edwards Deming

---

[35]See Footnote 28.

[36]W. Edwards Deming Quotes, URL: http://www.brainyquote.com/quotes/authors/w/w_edwards_deming.html#Vp4MldA51AqCIisT.99 (Access Date: 10 April 2014).

## Author Biography

**Tibor Babos** is the Project Manager of Hungary's First Critical Infrastructure Protection Scientific Research Program, he is a Honorary Professor at the Óbuda University, and at the National University of Public Service, Budapest, he is a Doctoral Thesis Advisor and Doctoral Supervisor at the University of Óbuda and at the National University of Public Service, Co-Chairman of the Transatlantic Policy Consortium, the Director of the National Security Center, Obuda University, as well as he is the Honorary President of the Postgraduates' International Network (π-net)

# Weather Forecasting System for the Unmanned Aircraft Systems (UAS) Missions with the Special Regard to Visibility Prediction, in Hungary

Zsolt Bottyán, Zoltán Tuba and András Zénó Gyöngyösi

**Abstract** Nowadays, Unmanned Aircraft Systems (UAS) systems are playing more and more significant role in military and civil operations in Hungary. The proper, detailed and significant meteorological support system is essential during the planning and executing phases of the different UAS missions. Within such systems, it is very important to generate an accurate short-time visibility prediction. In order to develop a mentioned short-time hybrid visibility forecast, we combined an analog statistical and a high-resolution WRF-based numerical predictions which, are available around the four main airports in Hungary. In our work we show the detailed construction of hybrid visibility forecast. To publish our results we created a special web site where the adequate meteorological predictions can be access in some graphical, text and other forms via (mobile) internet connection.

**Keywords** Unmanned aircraft system · Hybrid meteorological modeling · Integrated weather prediction system · Visibility prediction · Meteorological support

## 1 Introduction

Remote piloted aviation is highly sensitive to the real weather situation, because the dynamic processes of flight depends on the present state of the atmosphere. The atmospheric influence on the mentioned type of aviation is more important than the

Z. Bottyán (✉) · Z. Tuba
Institute of Military Aviation, National University of Public Service,
Szolnok, Hungary
e-mail: bottyan.zsolt@uni-nke.hu

Z. Tuba
e-mail: tubazoltan.met@gmail.com

A.Z. Gyöngyösi
Department of Meteorology, Eötvös Loránd University, Budapest, Hungary
e-mail: zeno@nimbus.elte.hu

© Springer International Publishing Switzerland 2016  23
L. Nádai and J. Padányi (eds.), *Critical Infrastructure Protection Research*,
Topics in Intelligent Engineering and Informatics 12,
DOI 10.1007/978-3-319-28091-2_2

normal flights one. Application of UAS systems, for both civilian and military purposes, is growing rapidly worldwide, due to lower operational costs of the airplanes, and these are going to decrease significantly in the near future [1]. The mentioned UAS systems are also playing more and more significant roles in military and civil operations in Hungary [2]. Aerial support for natural disaster management monitoring (earthquakes, floods and forest fire etc.), government and private survey (cartography, agriculture, wild life monitoring, border control, security and maintenance control for industrial companies, electricity cords network etc.) and defense of critical infrastructures may benefit from the on board instruments that might be the payload of such UAS's [3].

In spite of relatively easy control of most UAS systems, the weather hazards can be very dangerous for their flight like the manned ones, too. Despite the mentioned sensitivity of UAS systems to the dangerous weather phenomena—at present—there are few developed weather support systems for UAS handlers.

In order to decrease the weather-related risks during the UAS flights, we developed a complex meteorological support system for UAS users, mission specialists and decision makers. The mentioned system has to provide an adequate meteorological support during the planning and the flying phases of the UAS missions, with the special regard to the followings:

- Operational working during 0–24 h (always accessible)
- Readily accessible over the large geographical regions as far as possible, via internet connection (essentially everywhere, mainly via mobile net)
- Be simply installed in the other geographical regions (flexibility)
- In the final form it will contain a flight path optimization routine based on actual and predicted weather (future plan)

Finally, our developed meteorological support system can easily be implemented anywhere, since the applied meteorological data and numerical modeling system are mainly open-access!

## 2 Integrated UAS Weather Prediction System (IUWPS)

Forecasting the state of aviation meteorological variables is one of the greatest challenges for an operational forecaster [4]. For example, both visibility and ceiling (low clouds) are very difficult to predict even from the outputs of a very sophisticated high-resolution meteorological model.

Accordingly, the high-resolution numerical model output data should be processed parallel to statistical analysis of archive data base for a given weather situation and airports in order to yield the best forecast in a certain situation. This requires an integrated prediction system that consist both a suitable, specially tuned numerical model a statistical component, and—in addition—it is capable to generate an accurate hybrid (combined statistical and numerical) aviation meteorological forecasts.

There are some applied analog weather prediction procedures in an operational use, for example the Canadian fuzzy logic-based analog forecasting system called WIND-3 [5]. The success of the mentioned method gave us the motivation to develop a similar system in Hungary which is able to give effective meteorological support in UAS operations.

Visibility and ceilings play a key role in the success of UAS missions. Usually, the operational minimum of UAS flights is lower than the weather minimum of the special mission execution. For example, during reconnaissance or surveillance tasks, poor visibility and low ceilings can eliminate the mission but they do not restrict the UAS flight itself.

Our primary goal was to develop a hybrid forecasting system which is able to give accurate and timely forecasts of main airports with relevant climatic database to support UAS missions.

The construction of our experimental complex Integrated UAS Weather Prediction System (IUWPS) is based on the following parts:

- Statistical Modeling Subsystem (SMS)
- Numerical Modeling Subsystem (NMS)
- Hybrid Modeling Subsystem (HMS)
- Post Processing Subsystem (PPS)

The important components of the UAS meteorological system and its relations to their different components are shown on Fig. 1. The mentioned experimental IUWPS has a modeling and a post-processing unit using both statistical and numerical outputs of its subsystems to produce the hybrid forecasts in case of visibility and ceiling. The IUWPS uses climatological data of mentioned parameters from the Statistical Modeling Subsystem and actual weather forecast data (basic meteorological

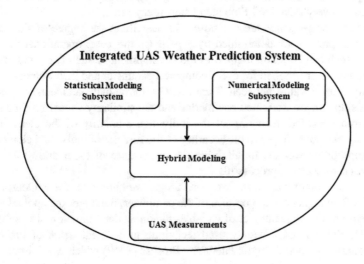

**Fig. 1** Structural chart of the integrated UAS weather prediction system

variables) from the Numerical Modeling Subsystem. On the basis of mentioned parts IUWPS is able to produce the hybrid (combined) short-time predictions regarding both visibility and ceiling.

## 3   Statistical Modeling Within the IUWPS

The basic principle of analog forecasting is to find similar weather situations in the past to the current and recent conditions and rank them according to the degree of their similarity in the interest of giving relevant information for weather forecasts. (The phrase of weather situation means couple of hour continuous observations in this study.) Therefore, analog forecasting does not work without a relevant climatic database which contains the meteorological parameters planned to forecast in the future. We have set up an appropriate database for Hungarian military airbases (LHKE, LHPA and LHSN) and for the largest Hungarian international airport (LHBP). It is based on routine aviation weather reports (METAR's) [6]. The database contains the meteorological variables for every half an hour from 2006. It has more than 30 variables, because it includes the parameters in elemental and derived format as well (e.g. year, month, day, hour, minute vs. Julian date). The records are more than 99 % complete. We have to note such database is easily creatable for any airport with available METAR weather reports all over the world, of course!

The applied fuzzy logic-based algorithm is measuring the similarity between the most recent conditions and the appropriate elements of the database. During the examination of every single weather situation the model uses the current and the eleven previous METARs' content. The algorithm compares the meteorological variables of every examined time step using fuzzy sets.

The fuzzy sets—which are composed for describing the degree of similarity— used in this process are determined by experts (in this case: operational meteorologists), which is a common method in the development of such fuzzy systems [7]. These functions are applied for every compared parameter and their outputs give the similarity with a value between 0 and 1. The individual parameter values of a weather situation are examined one by one and the similarity value of a time step is given from weighted averaging of the individual similarity of the elements [8]. Obviously, we could improve the accuracy of the forecast of individual elements by using appropriate weights highlighting the importance of them during the fuzzy logic-based forecasting process [8].

There is another approach, we can assign weighting to the meteorological variables. The higher the importance of the parameter, the larger the applied weight is. Because of the large number of variables the direct determination of weights was excluded. We applied a widely used technique in different fields of life except meteorology so-called Analytic Hierarchy Process (AHP) which was introduced by

Saaty [9]. This method is mainly used in multi-criteria decision making, especially in solving complex problems from most different fields [10, 11]. AHP was used only for determining the applicable weights for the different parameters as criteria. It is also necessary to apply pairwise comparison on criteria which is based on general definition. In our case these experts' judgments were assigned by operational forecasters' joint opinion. The ratios of pairwise comparisons can give the elements of a matrix. The best choice for the weight vector is the eigenvector belonging to the maximal eigenvalue of this matrix. For detailed description please refer Saaty's proof [8]. To determine the eigenvector we used the standard power iteration method. The received weights will be shown at the verification results. Obviously the matrix might be inconsistent due to the subjective comparisons. We found an inconsistency of 2.5 % which is less than the tolerable 10 %, so the results are significantly reliable [12].

# 4　Numerical Modeling Within the IUWPS

The Weather Research and Forecasting (WRF) model version 3.5 (release April 18, 2013) has been applied to generate numerical output for our general weather prediction system [13]. The WRF is a non-hydrostatic meso-scale meteorological model and its modularity and high flexibility with the wide global user experience suited well for the needs of our purposes. The modular structure of our development provides the possibility to exchange from one limited area model to another (e.g., HIRLAM/ALADIN, etc.) as a driver for the Numerical Modeling Subsystem unit of the system. Some successful experiments were conducted with CHAPEAU, the academic version of the ALADIN to verify the possibility of such replacement.

Input geographical data have been generated from two different sources. Land-cover/land-use information was taken from the CORINE 2000 (Coordinate Information on the Environment) database adapted and modified to be used in WRF [14]. The main advantage of this database with respect to the USGS data, originally used by WRF, is the much more realistic representation of land characteristic features (e.g., much better and more specified representation of various types of forests and shrub-lands; in addition to more than 3 times larger area specified as urban, etc.). These characteristics are essential in surface-atmosphere interactions and boundary layer processes, which are most important inputs for calculations of main aviation meteorology parameters.

Similarly, the original FAO (Food and Agriculture Organization) soil texture data has also been replaced by the DKSIS (Digital Kreybig Soil Information System, produced by the Centre for Agricultural Research, Hungarian Academy of Sciences) [15]. Contrarily to the CORINE database which can be applied to all model domains, the DKSIS covers only the area of Hungary so we were able to use it in the smallest (but highest-resolution) domain. In addition, soil hydraulic parameters were also adjusted according to Hungarian soil sample data (MARTHA and HUNSODA).

Within the whole prognostic area, which is located in the Carpathian-basin (centered at a location with geographical coordinates N47.43; E019.18), we applied a well-used three level telescopic nesting with 30 km horizontal resolution in the coarsest (d01) domain, 7.5 km in the intermediate (d02) domain and 1.875 km in the highest-resolution lowest level (d03) domain. The applied number of vertical levels was 44 and there were 24 levels in the lower troposphere under 2 km (Fig. 2).

In order to apply a setup tuned for the special requirements of the designated purpose, extensive tests were performed: 30 different combination of parameterization setups (ENS)—including 8 types of micro-physics (types 3–9 and 13), 6 types of Land Surface Models (types 1, 2, 4, 5, 7 and 10) and 8 types of PBL (1, 2, 4, 5, 7–10) schemes—have been tested. Tests have been performed for 9 selected weather situations in the model domain, which ones have strong aviation weather relevance (Table 1).

Model output has been compared to synoptic surface observations at 31 ground stations in Hungary and the operational radiosonde data of 4 stations located in the d02 domain. Temperature and dew point data have been evaluated using RMS error and the wind score derived with respect to wind speed and wind direction differences. Results showed that in the surface data there is a wide variation in humidity and temperature, while in the upper level data only wind speed and direction are significantly affected by the choice of the parameterization schemes. The final results of our parameter optimization procedure can be seen in Fig. 3.

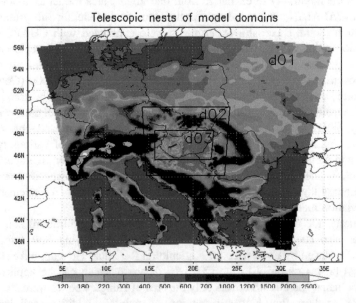

**Fig. 2** The applied telescopic nests of WRF model in the numerical modeling subsystem

**Table 1** The applied case studies and their dates during WRF optimization procedure

| # | Dates | Description |
|---|-------|-------------|
| 1 | 20121027 | Widespread precipitation from a Mediterranean low pressure system |
| 2 | 20120920 | High horizontal pressure gradient situation with strong winds, with a special wind-pattern |
| 3 | 20120119 | Significant low level inversion during winter period |
| 4 | 20120908 | High pressure ridge transition resulting in significant and rapid change in wind direction |
| 5 | 20120729 | Deep convection resulting in local and heavy showers that were not well resolved by most operational models |
| 6 | 20120512 | Significant change in wind direction following a cold front |
| 7 | 20120122 | Well documented severe icing case weather situation |
| 8 | 20120216 | Convective precipitation from a high level cold vortex, temperature in the mid-troposphere less than—25 °C |
| 9 | 20121206 | UAV test flight case for direct verification purpose |

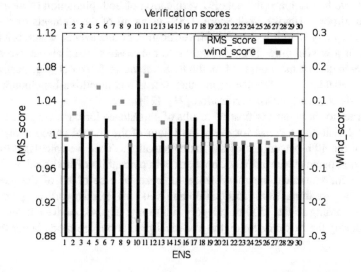

**Fig. 3** Final results of the parameterization optimization procedure of the WRF model. We have to note that in case of RMS scores the smaller values are the better ones, but in connection with wind scores the larger values are the better ones

On the basis of our analysis of the mentioned results, Bretherton and Park JC (9) PBL scheme [16], WSM Single-Moment 3-class (3) micro-physics scheme [17] and the Noah scheme [18] for land-surface processes were the best options. In addition, RRTM (Rapid Radiative Transfer Model) for long-wave radiation [19], Dudhia's scheme for short-wave radiation and a modified version of the Kain-Fritsch scheme for cumulus convection have been applied in the complete WRF model parameterization [20].

In our Numerical Modeling Subsystem 0.5° by 0.5° GFS data is applied as initial condition and boundary conditions for the limited area integration of the outermost domain every 3 h, with no additional data assimilation. Input data is preprocessed with WPS, the vendor preprocessor of the WRF system.

Integrations running for 96 h lead times are performed two times a day, initialized from 00 UTC and 12 UTC at 04:00 UTC and 16:00 UTC, respectively. Model products are delivered to the users through the web interface of the integration server itself 6 h after the observation of the input data.

## 5 Hybrid Forecasting of Visibility in IFS

As we mentioned earlier, the visibility and ceiling predictions are very difficult. In order to create well-used operational short-time (up to 9 h) forecasts we developed a hybrid (combined) method to predicting of these parameters for UAS users and specialists.

First, we had to make the detailed verification of both prediction of visibility at the four airports. Heidke Skill Score (HSS) using all of the elements of a special contingency table and it is correct with verification of rare events, which is a typical in case of low visibilities [21]. According to the reasons stated above, we present the HSS values of the examined methods. Some naive forecasts (e.g. persistence) can be a standard of reference or in other words a competitive benchmark in the field of short term forecast verification [21], [22].

In order to show our verification results of persistence forecast, we generated the statistical visibility forecast for every third hour of the control period applying the 30th and the 40th percentiles of the analog situations. Then we calculated the HSS verification parameter mentioned above for both percentiles of statistical predictions and also for numerical and persistence forecasts in case of all of the examined ICAO category limits (800, 1500, 3000 and 5000 m). In this work we present our results regarding to the 1500 m category limit only (Fig. 4). As we can see in this figure during the first nine hours the statistical predictions are much better than the

**Fig. 4** The HSS values as a function of prediction time of different visibility forecasts in case of 1500 m category limit

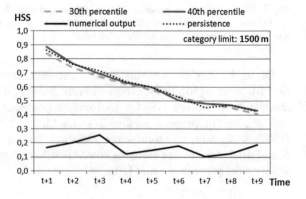

numerical one but the difference of accuracy between them are sharply decreasing after six hours. We gave very similar results in case of other examined category limits so this means the statistical predictions accuracy after nine hours not significantly better than numerical ones.

Applying our results mentioned above we created an experimental hybrid visibility forecast method based on our 40th percentile analog forecast with AHP weight and modified numerical WRF prediction. The construction of a hybrid forecast, is based on the followings:

$$VIS_{HYBRID} = a \cdot VIS_{STAT} + b \cdot VIS_{NUM} \qquad (1)$$

where $VIS_{HYBBID}$, $VIS_{STAT}$, and $VIS_{NUM}$ are the hybrid, the 40th percentile AHP analog and the WRF numerical visibility forecasts, respectively. The constants (a, b) are weights of the statistical and numerical forecasts. The applied values of the weight of $VIS_{STAT}$ (a) are found in a weight matrix represented by Table 2. The columns of this matrix are the time steps of forecasts (in hour) and its rows are the absolute category differences of the measured and numerical predicted visibilities at initial time (t + 0). If the difference value of the measured and numerically predicted visibility is within the 0–800 m, 800–1500 m, 1500–3000 m, 3000–5000 m and 5000–70,000 m intervals initially, the absolute category difference will be 0, 1, 2, 3 and 4, respectively. Obviously the weight b = 1 − a.

As it can be seen in the left hand side of Fig. 5 in the LHKE airport the measured (3000 m) and the numerical predicted visibilities (10,000 m) are highly different so the absolute category difference is 4 in this case. Contrarily in the right hand side of Fig. 5 the measured and numerical forecasted visibilities (10,000 m) at the LHSN airport are equal so the applied absolute category difference is 0. Accordingly the calculated $VIS_{HYBRID}$ forecasts are based on Table 2 and represented by the green columns (every column means one hour). Within the red rectangles we can also see the statistical (black line) and numerical (blue line) visibility forecasting based on Statistical and Numerical Modeling Subsystem of IUWPS. We have to note from t + 10 h the hybrid forecast will be the original numerical one because the accuracy of statistical predictions—by this time—is not enough to taking account of it to product more accuracy hybrid forecast (hence at t + 10 h the green column and black line show the same value of visibility).

**Table 2** The applied values of the weight of $VIS_{STAT}$ (a) statistical visibility prediction for hybrid forecasting

|  |  | t + 1 | t + 2 | t + 3 | t + 4 | t + 5 | t + 6 | t + 7 | t + 8 | t + 9 |
|---|---|---|---|---|---|---|---|---|---|---|
| Absolute category difference | 4 | 1.00 | 1.00 | 1.00 | 0.90 | 0.80 | 0.65 | 0.50 | 0.35 | 0.20 |
|  | 3 | 1.00 | 1.00 | 0.90 | 0.80 | 0.70 | 0.55 | 0.45 | 0.30 | 0.20 |
|  | 2 | 1.00 | 0.90 | 0.85 | 0.75 | 0.65 | 0.50 | 0.40 | 0.25 | 0.15 |
|  | 1 | 0.90 | 0.85 | 0.80 | 0.70 | 0.60 | 0.45 | 0.35 | 0.20 | 0.10 |
|  | 0 | 0.90 | 0.80 | 0.70 | 0.60 | 0.50 | 0.40 | 0.30 | 0.20 | 0.10 |

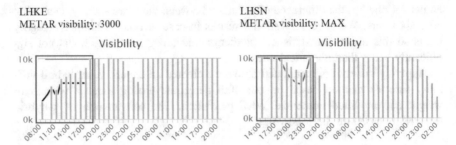

**Fig. 5** Visibility forecasts are calculated by IFS for two different initial situations. *Left* LHKE airport with measured 3000 m and numerical predicted 10,000 m visibilities. *Right* LHSN airport with measured 10,000 m and numerical predicted 10,000 m visibilities. Hybrid, statistical and numerical forecasts are represented by *green columns*, *black lines* and *blue lines*, respectively. The *red rectangle* shows the first ten forecasting hours. On the vertical axis 10 k represents 10,000 m visibility value

In order to permit of wide access of our weather predictions—with the special regards to hybrid visibility forecasts—we developed an experimental (in this time with restricted access) web-based weather information center for UAS users. It is a special web site, where the appropriate meteorological information will be accessed in a graphical, text and other formats via (mobile) internet connection in the future.

## 6  Summary

An accurate, detailed and significant meteorological support system is essential during the planning and execution phases of UAS missions. In such systems it is very important to generate an accurate short-time visibility prediction. The IUWPS is able to give this meteo support and is based on the followings main parts:

- An adequate data base of the main four Hungarian airports which contains the freely accessible METAR data (this data base can be extensible similarly)
- The applied analog statistical model with AHP weights can help to give accurate prognostic information about visibility for the first nine hours (Statistical Modeling Subsystem)
- The WRF based numerical weather model can give us high resolution complex weather prediction (of course also the visibility forecasts) over the Hungary with the maximum horizontal resolution of 1.8 km (Numerical Modeling Subsystem)
- The applied post-processing methods—are based on the mentioned statistical and numerical visibility products—to predict the combined hybrid visibility forecast for the first nine hours at the region of four Hungarian airports (Hybrid Modeling Subsystem)

- A special web site where the adequate meteorological information can be access in some graphical, text and other formats via (mobile) internet connection (Post Processing Subsystem)

In the future we would like to develop an UAS flight path optimization, which is based on our predicted 3D weather situation and the planned UAS flight. We will also continue the development and testing of our Hungarian Unmanned Meteorological Aircraft System (HUMAS) which will be able to help verify and improve our weather prediction capability.

**Acknowledgements** This work was supported by the European Social Fund (TÁMOP-4.2.1. B-11/2/KMR-2011-0001, Critical Infrastructure Protection Research).

# References

1. Gertler, J.: U.S. Unmanned aerial systems. Congressional Research Service. https://www.fas. org/sgp/crs/natsec/R42136.pdf (2012)
2. Fekete, Cs., Palik, M.: Introduction of the hungarian unmanned aerial vehicle operator's training course. Defense Resour. Manag. 21st Century **1**, 55–68 (2012)
3. Gyöngyösi, A.Z., Kardos, P., Kurunczi, R., Bottyán, Z.: Development of a complex dynamical modeling system for the meteorological support of unmanned aerial operation in Hungary. In: Proceedings of International Conference on Unmanned Aerial Systems, pp. 8–16. Atlanta, USA, IEEE (2013)
4. Jacobs, A.J.M., Maat, N.: Numerical guidance methods for decision support in aviation meteorological forecasting. Weather Forecast **20**, 82–100 (2005)
5. Hansen, B.K.: A fuzzy logic-based analog forecasting system for ceiling and visibility. Weather Forecast **22**, 1319–1330 (2007)
6. Bottyán, Z., Wantuch, F., Gyöngyösi, A.Z., Tuba, Z., Hadobács, K., Kardos, P., Kurunczi, R.: Development of a complex meteorological support system for UAVs. World Acad. Sci. Eng. Technol. **76**, 1124–1129 (2013)
7. Meyer, M.A., Butterfield, K.B., Murray, W.S., Smith, R.E., Booker, J.M.: Guidelines for eliciting expert judgment as probabilities or fuzzy logic. In: Ross, T.J., Booker, J.M., Parkinson, W.J. (eds.) Fuzzy Logic and Probability Applications: Bridging the Gap, pp. 105–123. Society for Industrial and Applied Mathematics (2002)
8. Tuba, Z., Vidnyánszky, Z., Bottyán, Z., Wantuch, F., Hadobács, K.: Application of analytic hierarchy process (AHP) in fuzzy logic-based meteorological support system of unmanned aerial vehicles. Acad. Appl. Res. Mil. Sci. **12**, 221–228 (2013)
9. Saaty, T.L.: A scaling method for priorities in hierarchical structures. J. Math. Psychol. **15**, 234–281 (1977)
10. Gyarmati, J., Felházi, S., Kende, G.: Choosing the optimal mortar for an infantry battalion's mortar battery with analytic hierarchy process using multivariate statistics. In: Decision Support Methodologies for Acquisition of Military Equipment Conference, Bruxelles, NATO RTO, pp. 1–12 (2009)
11. Al-Harbi, K.M.: Application of the AHP in project management. Int. J. Project Manage. **19**, 19–27 (2001)
12. Saaty, T.L.: Some mathematical concepts of analytic hierarchy process. Behaviormetrika **29**, 1–9 (1991)

13. Skamarock, W.C., Klemp, J.B., Dudhia, J., Gill, D.O., Barker, D.M., Duda, M.G., Huang, X.-Y., Wang, W., Powers, J.G.: A Description of the Advanced Research WRF Version 3 NCAR/TN–475 + STR, NCAR Techical Note. (2008)
14. Druszler, A.: Meteorological effects of the land cover types changes during the twentieth century in Hungary. PhD Thesis, University of West Hungary, Sopron, p. 137 (2011)
15. Pasztor, L., Szabo, J., Bakacsi, Zs: Digital processing and upgrading of legacy data collected during the 1:25.000 scale Kreybig soil survey. Acta Geodaet. Geophys. Hung. **45**, 127–136 (2010)
16. Bretherton, C.S., Park, S.: A new moist turbulence parameterization in the community atmosphere model. J. Clim. **22**, 3422–3448 (2009)
17. Hong, S.-Y., Dudhia, J., Chen, S.-H.: A revised approach to ice microphysical processes for the bulk parameterization of clouds and precipitation. Mon. Weather Rev. **132**, 103–120 (2004)
18. Chen, F.J., Dudhia, J.: Coupling and advanced land surface-hydrology model with the Penn State-NCAR MM5 modeling system. Part I. model implementation and sensitivity. Mon. Weather Rev. **129**, 569–585 (2001)
19. Mlawer, E.J., Taubman, S.J., Brown, P.D., Iacono, M.J., Clough, S.A.: Radiative transfer for inhomogeneous atmospheres: RRTM, a validated correlated-k model for the longwave. J. Geophys. Res. **102**, 16663–16682 (1997)
20. Kain, J.S.: The Kain-Fritsch convective parameterization: an update. J. Appl. Meteorol. **43**, 170–181 (2004)
21. Doswell III, C.A., Davies-Jones, R., Keller, D.L.: On summary measures of skill in rare event forecasting based on contingency tables. Weather Forecast. **5**, 576–585 (1990)
22. Murphy, A.H.: Climatology, persistence, and their linear combination as standards of reference in skill scores. Weather Forecast. **7**, 692–698 (1992)

# Tribological Aspects for Reliable Operation of Engineering Surfaces

Árpád Czifra and István Barányi

**Abstract** Reliable operation of machine elements depends on the tribological properties of contacting elements, such as, the quality of the surface. Microgeometrical characterization of engineering surfaces has standard and non-standard methods, but in many cases of worn surfaces the results of these techniques become unusable. After examining worn surfaces—differential of military land vehicle, cylinder liner of engines and ferrodo-steel sliding couple—a new concept of surface characterization was developed: the feature based concept of surface characterization focusing to the properties of microtopography which determine the operation.

**Keywords** Surface features · Reliability · Wear

## 1 Introduction

Besides dimensional accuracy and shape accuracy, the operation, reliability and life cycle of components produced in different ways highly depend on the quality of the surface microgeometry.

Traditionally and in accordance with Hungarian and international standards, the microgeometry of operating surfaces has been characterized by two dimensions; however its information content is limited. Thomas and Rosén [1] proved that parameter-based techniques are uncertain and limited in many ways, because the results greatly depend on evaluation length, sampling distance and filtering. Nowadays—beyond the parameter based technique—two dominant research trends can be observed in surface roughness characterization. One is the technique when the local features of topographies are characterized based on the identification of

Á. Czifra (✉) · I. Barányi
Donát Bánki Faculty of Mechanical and Safety Engineering, Óbuda University,
Budapest, Bécsi út 96/B 1034, Hungary
e-mail: arpad.czifra@bgk.uni-obuda.hu

© Springer International Publishing Switzerland 2016
L. Nádai and J. Padányi (eds.), *Critical Infrastructure Protection Research*,
Topics in Intelligent Engineering and Informatics 12,
DOI 10.1007/978-3-319-28091-2_3

asperities and scratches [2], while the other is the "global" surface characterization method using complex mathematical tools [3].

Both production engineers and tribologists use standard and non-standard methods to characterize engineering surfaces. It is well known, that the average surface roughness (Ra) and surface height (Rz) preferred and widely used in the industries can only restrictively characterize the engineering surfaces. At the same time the rapid development of manufacturing and material technology keeps raising the need that a correlation could be made—in terms of the basic surface roughness parameters—between the quality of cut surfaces and cutting parameters. Horváth et al. [4] dealt with the cutting capacity of aluminum alloys and with the cutting abilities of diamond tools. Phenomenological models were built to estimate Ra and Rz within the examined range of cutting parameters [5, 6] extent these examinations to milling and drilling processes. All these studies can be important to in terms of applying the particular materials and technologies in different industries; however, these examinations investigations do not give enough information on the functional properties of the manufactured surface. Tóth et al. [7] performed complex characterization of machined surfaces to analyze the integrity of differing surface micro-geometries with the measurements of surface parameters, as well as with the calculation and discussion of statistical features of the results.

In this study authors point that operating surfaces—especially in case of wear—need special characterization method to find information between surface and tribological properties. A new concept for surface microgeoemetry analysis is presented here using standard and non standard methods.

## 2  Classical and New Concept of Surface Characterization

Although different techniques are used to characterize surface microtopography the basic concept of the analysis is very similar. Figure 1 show the classical method of surface roughness characterization. In that case different production and operational

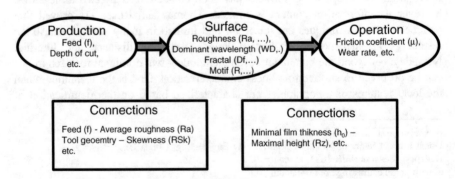

**Fig. 1** Classical concept of surface characterization

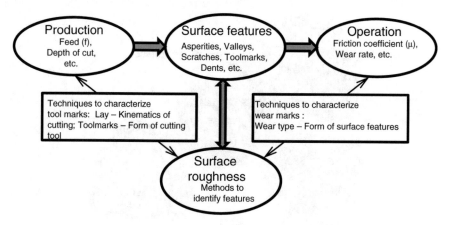

**Fig. 2** Feature based concept of surface characterization

properties are connected to surface roughness parameters. Main problem of this concept is that surface parameters have limited information content and cannot represent the surface microgeometry. If some parameter in machining or operating process changes (tool wear, abrasion, etc.) absolutely different surface appear.

The local features of surfaces—such as asperities, scratches, valleys, dents, toolmarks, scuffmarks, furrows, nicks, etc.—are determine the tribological behavior. The aim of this study is to introduce a new view-point in surface roughness characterization. The goal is to find a method for feature-based characterization that can identify the dominant elements (features) of the surface and can characterize them. Figure 2 represent the philosophy of feature-based surface evaluation.

Following chapters summarize some results of analysis of worn surfaces to present the motivations and methods how and why to use feature-based surface characterization.

# 3  Analysis of Worn Surfaces

## 3.1  Differential of a Military Land Vehicle

Machine elements of military vehicles are often working under high loads and in special circumstances. The investigated differential came out from a vehicle used in Iraq. Construction of differential and investigated element can be seen in Fig. 3.

Connecting parts are in Hertz-contact. Due to high load severe wear has been occurred. Based on micro hardness and sectional analysis it was proved that very hard layer was formed on surface. The high loads causes micro-cracks on this layer and related damage. Figure 4 shows scanning electron-microscopic (SEM) and topographic image from surface.

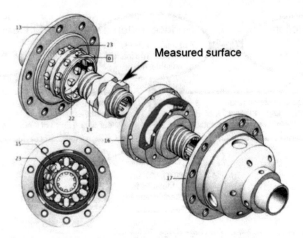

**Fig. 3** Investigated machine element of differential

**Fig. 4** SEM image and microtopography ($z_{min} = -29.4$ μm; $z_{max} = 5.1$ μm) of differential element

Beside tribological analysis, surface roughness measurements were carried out. On differential machine element $2 \times 4 \times 10$ profile were measured. In four surface 10–10 tangential and axial standard measurement with 5.6 mm measuring length and 0.8 mm cut-off were carried out. The measuring system was Mahr Perthometer, with MFW-250 stylus.

Figure 5 shows that worn surface has a plateau like peak zone and some deep valley in valley zone. Density of these deep valleys are different (some profile does not contain any one). Table 1 summarizes some parameters of profiles. Standard deviation and range of parameters are so high that classical statistical characterization with parameters has no interest, although similarities of profiles can be observed.

It is well known, that a correlation can be observed between some parameters. For instance $Rz = 4.5\ Ra$. It is also known that these relations can change depending on machining technique. For example $Rz = 4.5\ Ra$ in case of turning, but $Rz = 7\ Ra$

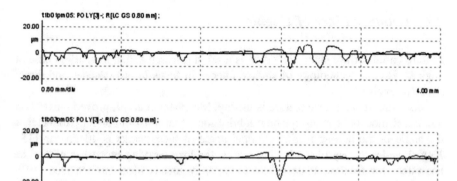

**Fig. 5** Roughness profiles of differential

**Table 1** Parameters of worn surface of element of differential

| Parameter | Ra (μm) | Rz (μm) | Rv (μm) | RSm (μm) | RSk (−) |
|-----------|---------|---------|---------|----------|---------|
| Average   | 1.25    | 7.19    | 4.52    | 339      | −1.14   |
| Stan. dev.| 0.82    | 4.09    | 2.47    | 162      | 1.15    |
| Range     | 4.01    | 18.6    | 10.1    | 1164     | 6.03    |

in grinding (see [8]). So the relationship of parameters can represent the surface features.

The relationship between amplitude parameters—such as the ratio of Rz and Rv, or Rv and Rp—represents the main features of surfaces. Figure 6 shows the correlation between Rv and Rz parameters in case of worn differential machine element. Based on statistical results (Table 1) we supposed that no correlation can be found, but Fig. 6 shows strong correlation between Rz and Rv.

A strong correlation means that same tribological circumstances form similar surfaces; however the roughness of these surfaces can alter.

**Fig. 6** Correlation between Rz and Rv parameters of differential machine element

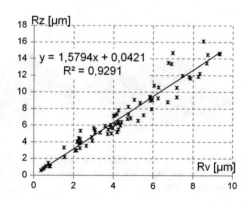

## 3.2  Cylinder Liner of Engine

Feature based surface characterization is used in research and in industry in case of engines. Honing technology of surface liner is in focus in last decades and directives are evolved.

Most important surface feature is the lay. Mezghani et al. [9] proved—based on numerical models of hydrodynamic lubrication—that most important parameter is the angle of the honing marks. Optimal angle of these marks is 40–55° or 115–130°. Besides the lay of the honing marks, the load carrying properties of surface has effect for friction. Plateau like topography means ideal surface (see [10]). Examinations contained different surface roughness parameters, but only week correlation was experienced in case of Rpk, Rvk and Rk parameters. It confirm that feature based surface characterization is more effective than classical way of surface evaluation.

New trends of development and design of surface of cylinder liner is the modification of surface with laser texturing (see Fig. 7). These features influence the lubricating properties of surfaces and results beneficial behavior even in case of boundary lubrication (see [11]).

It is an important question as to how this well-designed surface changes, due to wear. Worn surface liner was measured to identify topographic features after wear.

On Fig. 8 worn topography can be seen. The original honing marks fully disappeared and scratches form in direction of relative movement. Somewhere small pits and nicks can be observed (see Fig. 8 A and B point). Depth of these pits is about 10 μm or in some cases they are deeper. Figure 8 sign C shows a deep scratch that may formed by hard particle. The surface has a plateau like peak zone; high negative values of SSK parameter demonstrate this phenomenon (surface skewness, SSK parameter is in a range −2.1 and −7.3).

Altogether the main reason of the analysis is that the plateau like surface is an optimal formation because it provides excellent load carrying and this surface feature become "permanent" in wear. The original honing marks are disappear in wear and new surface has not got optimal retention of lubrication, so deeper laser marks can provide better lubrication in long term operation.

**Fig. 7** Laser textured surface of cylinder liner (from *left* to *right* 4 × 4 mm topography; 250 × 250 μm topography with 20 μm depth and 60 μm width laser mark; scanning electron microscopy image)

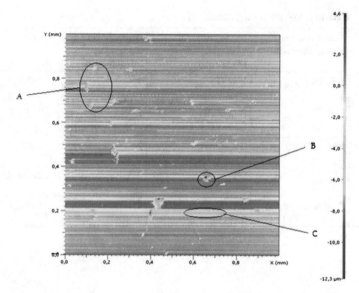

**Fig. 8** Worn surface (1 × 1 mm topography, 2 × 2 μm sampling) of cylinder liner

## 3.3  Characterization of Abrasive Wear Process

Deeper analysis of surface changes in wear provides possibility to find main features and feature-based methods. In present part of study topographical measurements and special 3D parameter-based characterizations were performed to study the changes of microtopographies in abrasive wear. Laboratory wear tests of ferrodo-steel sliding pair were carried out and topographies before and after wear were investigated. Fractal dimension (Df), root mean square slope of surface (Sdq) of topographies before and after wear were analyzed and compared.

Steel (55Si7) sliding tracks were used for the tests, on which Ø 18 mm ferodo specimens were slid. Two different grinding wheels were used for producing 'fine' (sign: F) and 'rough' (sign: R) surfaces. Two machining directions were used: one of them was perpendicular to (sign: N) while the other was parallel (sign: P) with the sliding direction. The stroke of the sliding was 180 mm and the load was 1000 N. The sliding speed was 50 mm/s. No lubrication was applied. First test series were slide for 2 h while the second ones were slide for 6 h. Steel track microtopography was recorded at two different places using a Mahr Perthen 3D type stylus instrument. The size of the measured surface was 1 × 1 mm with 2 μm sampling in both directions. Measurements were performed on identical surface sections before and after the wear process; therefore the changes of a given surface section can be traced accurately, not only statistically. Table 2 summarizes set-up of topographic measurements.

Figures 9 and 10 show some topographies. The original microtopography of 2hNF_E surface disappeared and new surface texture was formed in accordance

**Table 2** Wear tests and measuring conditions

| Wear time | Machining versus sliding direction | Average roughness of original topography | Repetitions of topography measurements | Sign |
|---|---|---|---|---|
| 2 h | Parallel (P) | Fine (F) Sa ≅ 0.2 μm | 2 | 2hPF_A |
| | | | | 2hPF_B |
| | | Rough (R) Sa ≅ 1.6 μm | 2 | 2hPR_C |
| | | | | 2hPR_D |
| | Perpendicular (N) | (R) Sa ≅ 1.6 μm | 2 | 2hNF_E..F |
| | | (F) Sa ≅ 0.2 μm | 2 | 2hNR_G..H |
| 6 h | Perpendicular (N) | (F) Sa ≅ 0.1 μm | 6 | 6hNF_I..N |

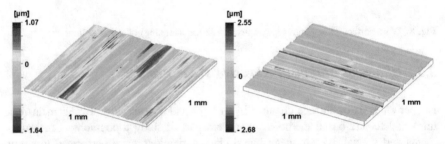

**Fig. 9** 2hNF_E surface before wear and after 2 h sliding

with the direction of relative movement. The scratches formed seem "fine" and uniform. In the case of Fig. 10, the original surface texture has not disappeared, but the peak zone of microtopography changed: scratches were formed in the sliding direction. Scratches are long on the surfaces and not too deep, so abrasive mild wear can be supposed in all cases.

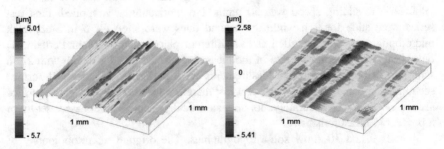

**Fig. 10** 2hNR_H surface before wear and after 2 h sliding

Characterization of topographies based on ISO 25178 standard parameters and on fractal dimension analysis [12]. Only two parameter is analyzed in details: Sdq —RMS gradient (average slope of surface) and Df—fractal dimension.

Surface parameter Sdq is height-independent parameter and characterizes the average slope of surface. It means that Sdq value characterizes surface features well than average roughness and other amplitude parameters. Figure 11 shows changes of this parameter in wear. It can be observed that quality of surface has no effect to the parameter: in case of A, B, G, H (fine surface with Sa = 0.2 μm) Sdq = 2.5°, C, D, G, H (rough surface with Sa = 1.6 μm) Sdq is higher than 10°, but in case of "very fine" surface (I...N, Sa = 0.1 μm) Sdq is about 8°. During wear Sdq changes dramatically. Some cases the change is higher than 100 %. In cases of A, B, G, H the parameter increases, although the wear processes of A–H specimens are the same. The Sdq value after wear—using similar test set up—becomes similar. In case of A–H this value is in range 3.62–7.08°, although the original topographies have wide range (2.73–12.76°) of this parameter. Analyzing the results of six-hour wear process the conclusion is the same.

Fractal dimension of topographies were calculated with 3D power spectral density analysis. The result of a 3D analysis is a surface in the frequency domain. In logarithmic scale, frequency-PSD amplitude can be approximated by a straight line. The slope of the line is in correlation with the fractal dimension. Details of the method are in [13]. Figure 12 shows the fractal dimension before and after wear. Df decreased in examined wear processes, except three cases, where Df increased. Worn topographies in these cases have deep groves formed by wear particles, and these features are causes the increases of Df.

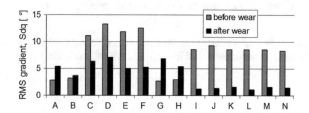

**Fig. 11** RMS slope before and after wear

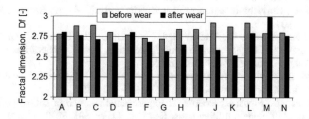

**Fig. 12** Fractal dimension before and after wear

# 4 Conclusion

Information obtained from the micro and nano topographies, of operating surfaces, appears as input in current machining and tribological models. The reliability of these models is in close correlation with the effectiveness of topographical analyses. Tribological processes can only be understood and accurately modeled, through an in-depth knowledge of surface topography.

Classical methods to characterize engineering surfaces are uncertain and limited in many ways, because the results are not in connection with main topographic elements. Therefore the authors suggest new viewpoint in surface analysis: a feature based method.

The following conclusions are proven herein:

- Profile parameters have height standard deviation for worn surfaces, so these parameters are not suitable to characterize the microgeometry. In analyzed cases, strong correlation of Rz and Rv parameter were found, which means that worn surfaces under the same operating conditions, have similar topographic features.
- Examination of cylinder liner confirmed that plateau like topography is beneficial in operation, but due to wear original honing marks are disappear. It means that deeper surface features (such as laser marks) are needed to provide better lubrication in a long term operation.
- In an abrasive wear process the Sdq parameter has changed in accordance of operating conditions: similar operating conditions results same Sdq values. Fractal dimension characterization of topographies based on power spectral density provides an opportunity to investigate wear processes.

**Acknowledgements** We acknowledge the financial support of this work by the Hungarian State and the European Union under the TÁMOP-4.2.1B-11/2/KMR-2011-0001 project.

# References

1. Thomas, T.R., Rosén, B.G.: Determination of the sampling interval for rough contact mechanics. Tribol. Int. **33**, 601–610 (2000)
2. Czifra, Váradi: Horváth: three dimensional asperity analysis of worn surfaces. Meccanica **43**, 601–609 (2008)
3. Persson, B.N.J., Albohr, O., Trataglino, U., Volokitin, A.I., Tosatti, E.: On the nature of surface roughness with application to contact mechanics, sealing, rubber friction and adhesion. J. Phys. Condens. Matter **17**, R1–R62 (2005)
4. Horváth, R., Drégelyi-Kiss, Á.: Mátyási Gy: Application of RSM method for the examination of diamond tools. Acta Polytech. Hung. **11**(2), 137–147 (2014)
5. Palásti-Kovács, B., Sipos, S., Szalóki, I.: Experimental research of cutting performance and quality abilities of modern drilling tools. Key Eng. Mater. **581**, 32–37 (2014)
6. Mikó, B., Beno, J., Mankova, I.: Experimental verification of cusp heights when 3D milling rounded surfaces. Acta Polytech. Hung. **9**(6), 101–116 (2012)

7. Tóth, G.N., Drégelyi-Kiss, Á., Palásti-Kovács, B.: Analysis of microgeometrical parameters of cutting surfaces. Pollack Periodica **8**(2), 55–66 (2013)
8. Palásti-Kovács, B., Sipos, S., Czifra, Á.: Interpretation of "Rz = 4 × Ra" and other roughness parameters in the evaluation of machined surfaces. In: Proceedings of the 13th International Conference on Tools, ICT, Miskolc, 27–28 Mar 2012, pp. 237–244 (2012)
9. Mezghani, S., Demirci, I., Zahouani, H., El Mansori, M.: The effect of groove texture patterns on piston-ring pack friction. Precis. Eng. **36**(2), 210–217 (2012)
10. Grabon, W., Pawlus, P., Sep, J.: Tribological characteristics of one-process and two-process cylinder liner honed surfaces under reciprocating sliding conditions. Ribology Int. **43**(10), 1882–1892 (2010)
11. Keller, J., Fridrici, V., Kapsa, Ph, Huard, J.F.: Surface topography of cast ironin boundary lubrication. Tribol. Int. **42**, 1011–1018 (2009)
12. Barányi, I., Czifra, Á., Horváth, S.: Power spectral density (PSD) analysis of worn surfaces. In: Gépészet 2010 Proceedings of the Seventh Conference on Mechanical Engineering, Budapest, 2010. május 25–26, ISBN 978-963-313-007-0 (2010)
13. Czifra, Á., Goda, T., Garbayo, E.: Surface characterisation by parameter-based technique, slicing method and PSD analysis. Measurement **44**(5), 906–916 (2011)

# Human Factor Analysis in Unmanned Aerial Vehicle (UAV) Operations

## Aeromedical and Physical Approach

Zoltán Dudás, Ágoston Restas, Sándor Szabó, Károly Domján
and Pál Dunai

**Abstract** In Hungary Unmanned Aerial Vehicle (UAV) studies have a new momentum from a research project on Critical Infrastructure Protection (CIP). A Hungarian research team from the National University of Public Service made an attempt to address the UAV's human factor related safety issues, as selection, training and licensing. In UAV operations the achievement of an operational task is highly dependent on the operators' proper physical and mental performance, sensomotoric skills and abilities, so a medical screening shall be incorporated into the general selection process. The medical requirements naturally can differ from the physiological parameters of the pilot performing real flight because the weighted importance of aeromedical stressors and the level of flight safety risk are different. The evaluation of mental and physical performance and assessment of stress tolerance might be crucial from flight safety and operational aspects. The importance of simultaneous evaluation of cognitive performance, accompanying psychic stress and vegetative stress indicators is also emphasized.

**Keywords** Flight safety · Critical infrastructure protection · Remotely piloted aviation · Aircrew training · Psychic stress tolerance · Physiological parameters · Heart rate variability · Flight simulator

Z. Dudás (✉) · Á. Restas · S. Szabó · K. Domján · P. Dunai
National University of Public Service, Budapest, Hungary
e-mail: doktorpilot@gmail.com

Á. Restas
e-mail: restas.agoston@uni-nke.hu

S. Szabó
e-mail: sasi19620@gmail.com

K. Domján
e-mail: domjan.karoly74@gmail.com

P. Dunai
e-mail: Dunai.Pal@uni-nke.hu

© Springer International Publishing Switzerland 2016     47
L. Nádai and J. Padányi (eds.), *Critical Infrastructure Protection Research*,
Topics in Intelligent Engineering and Informatics 12,
DOI 10.1007/978-3-319-28091-2_4

# 1 Introduction

The UAV proliferation these days is not a question. UAV's beside their military applications are obvious to use as CIP elements for example political and state administration centers, hazardous facilities, power plants, important economical facilities etc. Though developing UAV operations for state purposes on one hand can be very effective, the other hand for instance their safely integration in the conventional air traffic raises many questions. Its complexity and the rate of its permanent development and growing diversity of UAV operations are now projecting some safety conflicts with the conventionally piloted aviation system in the near future. A drone flying on a surveillance mission over a place crowded with people or over a nuclear facility could be both a defense element and even a threat, when something goes wrong. A crashing UAV can weigh over a 150 kg and has the kinetic energy to take lives and ruin vulnerable elements of the state infrastructure.

Flight safety is traditionally based on three factors: as on technical factors (aircraft, systems, airports, maintenance etc.), environmental factors (weather, natural phenomenon) and human factors, which are the more important of all.

Besides the technical development there are great potentials and possibilities in the development of human elements of the UAS. The selection, training, and licensing of UAV crews so far have not been fully covered by the national law. Though, their scientifically based definition would be a good token of operational safety. That is why our research is headed to a solution of ending those deficiencies.

Safety, even in the case of UAV flight, still lies in human hands. From a small harmless model up to a globetrotter combat UAV regardless to their control method cannot be flown without a certain rate of human control. The main decisions during missions are made by a human element. Finally, let it be a direct controlled or just overseen UAV mission the safety itself finally are guaranteed by the aircrew. Without them the most sophisticated technical systems are just useless having no decision making capability during flight and in emergency. The ability, quality of the human element means the basis of safety.

Therefore, the aircrews were put in the focus of studies, by using the following hypothesizes:

- UAV mission safety is basically determined by human elements of the system; UAV pilots should be considered as full pilots; and they may have overlapping in competencies with other professionals within the aviation domain; UAV crews should be differentially licensed according to aircraft category, type and mission.
- Among the physical skills, in addition to the basic level of conditional skills, complex coordination skills play a defining role in the skill development of UAV operators.
- In coordination and psychomotor skills tests, pilots perform better than the average of other control group.
- Stress tolerance might be used as a parameter for UAV operator working ability: good operator might have a better stress tolerance, solving the task at lower

stress level. Positive correlation is supposed between favorable (closer to normal range) physiological parameters, better results in standard psychological instrument tests and better performance in simulated flight tasks on ground-based simulator. In this case, the subject is capable to perform the same task at lower stress level (possibly in real UAV mission as well).

The questions for which our studies try to find the answers are the following:

From human factors' aspect could it be possible to provide a same level of safety as conventional flights during UAV operations? Could a UAV flight be as safe as a conventional flight while the technical, personnel requirements and rules—if they even exist—are far more forgiving than those for conventional ones?

## 2   Research Methods

During our studies on UAV HF, conventional research methods seemed to be the most appropriate, like document analysis (reports, monographies, articles, statistics), critical adaptation (existing operational procedures) and personal interviews with UAV personnel. As for the HF studies taking the existing technical level (safety, reliability) theoretically constant and excluding a changing element of the system seemed to be a feasible simplification. While different characteristics of UAV categories were taken into account. Likewise, the environment is considered as still and limited (horizontally and vertically limited, appropriate weather conditions). This way HF could be taken under a deep scrutiny without being disturbed by volatile elements of the system.

The research was conducted in different directions as:

- UAV crew member selection protocol (medical, suitability for aircraft control, pre-qualification)
- UAV crew training minima and optima according to mission (military, police, disaster control) and aircraft category (VLoS-BLoS, control method)
- Crew Resource Management  of UAV
- Simulations in selection and UAV crew training
- UAV crew licensing criteria (medical, qualification, level of proficiency)

## 3   Using a Simulator in Experiments, Training and Selection

The simulator console used during our tests is a self-designed one. Its development was led by the aspects of the research. The idea of having a simulator console originates from the need for testing persons and measuring their abilities during the selection phase. The simulator is also suitable for observation their development in

UAV control even in VLOS (Visual Line of Sight) and BLOS (Beyond Visual Line of Sight) mode. Meeting these needs, the team decided on utilization of simple RC training software for external pilot view and flight simulator software for internal pilot view. The console itself is also a central factor in the medical tests when the subject persons' physiological functions (heart rate, saturation, skin resistivity etc.) are measured and recorded during simulated emergency situation. Since simulators usually are stationary and expensive, the research group decided on a different approach providing flexibility in usage and cost efficiency. It contains a high performance notebook, with two pieces of simulation software installed and with special controls for different control modes, such as direct internal and external pilot control. This configuration also provides flexibility in uploading several RPA types to the system. The file system is structured the way that special files can be up or downloaded from a remote server. So the software configuration is configurable in a flexible way and the flight data is obtainable from the remote. The simulator training for RPA pilots this way can be conducted in a virtual airspace without the need for the presence of an instructor. The RPA simulator training is also supported by a training syllabus with comprehensive training material and a multimedia interactive handbook.

During standardized flight missions on a simulator we have measured the pulse variability in medium frequency domain in order to characterize the psychosomatic stress level and its alteration in critical phases of flight. In the final protocol specific tasks (take off and landings, en-route flight) were measured, giving a profile about the undulation of stress level and responsiveness, providing an impression about psychic stability. We have measured the performance of:

(a)  Real UAV operators with operational experiences
(b)  Rotary wing pilots from an Air Force Base in Hungary
(c)  Young students from secondary school (experienced in video simulator games such as car race games)

The other main direction for our HF research was practical training. Among limited financial conditions efficiency is always a keyword when expensive technical systems like UAS are operated. So as to replace a part of the practical flight training by a zero risk simulator training seems to be the perfect solution for risk mitigation. Using this notion, our team worked out a simulator training syllabus and of course an exact evaluation system for UAV pilots. The job was difficult because at the beginning we did not possess any accurate measuring methods for the assessment of the candidates' performance. Once they were in place, the results became comparable while the training minima and optima definable. During our simulations we use a control group for the selection and training processes' test. There were some debates among our experts about the composition of the experimental subjects' group. Some advised that only pilots from different branches should be tested once the group decided to consider UAV pilot as conventional pilots. Others insisted on using air traffic controllers since their competencies' overlapping with other aviation professionals. Finally, a decision was made on having both pilots and ex-pilots with different backgrounds, and flight traffic controllers and aircraft modelers included. Later, we enlarged the control group with computer fan teenagers.

## 4 Pulse Variability Measurements to Assess Psychosomatic Stress Level During Simulated Flights

By definition the UAV (Unmanned Aerial Vehicle) or RPA (Remotely Piloted Aircraft) is not under the personal control of the pilot onboard, but regarding as a complex combat flight system the responsibility of the operator in the ground based control station is crucial. The process of medical screening of operators should be similar to the selection of pilot candidates, but surely less expensive: the narrowed spectrum of medical examination (e.g. no sense to perform barochamber and centrifuge training), the extended utilization of flight simulators contribute to the overall economic application of UAV systems. Instead of threats posed by classic aero-medical stressors (hypoxia or G overloads) the main problem is loss of spatial orientation regarding UAV position and attitude. In order to avoid loss of Situational Awareness, a high level of mental and sensorial performance is a must. From flight safety aspect the sudden incapacitation could commence at the UAV operators most likely as a mental breakdown or frozen mental state which could lead easily to the loss of the "Big Picture". The American Aerospace Medical Association, recently focused on human factor problems of UAV operators [1].

At each subject we monitored the (instantaneous) heart rate in experimental settings using ECG-like electrodes and recorded the beat-to-beat (RR) heart cycles by ISAX (Integrated System for Ambulatory Cardio-Respiratory Data Acquisition and Spectral Analysis, prototype 5.0) [2]. The quasi-periodic changes in electro mechanic cycles of heart summarized in a specific way (super positioning) resulting in variability. In short periods (second and minute ranges) these undulations can be divided in 3 main domains using mathematical models [3, 4]. These fluctuations in high frequency domain related to respiration, but in medium frequency domain (around 0.1 Hz) the increased rigidity (lowered fluctuations) is a consequence of vegetative imbalance, increased stress even for mental workload [5]. It is also stated, that after the suppression of oscillation there might be a rebound, which magnitude is proportional to the previous workload. For longer continuous workload the suppression undulation in time elapsed (deviation) is proportional to the workload [6]. We have examined the intensity changes in Medium Frequency domain (P_MF spectral or Power average) at rest and during different phases of simulated flight in similar and comparable time-periods, calculated the deviations before, during and after the flight. The raw files (containing RR beat-to-beat interval data) were saved on laptop, later on transformed into *psd* files. Based on log markers the comparable periods were identified and analyzing software fast Fourier transformation was executed, screening out the oscillation in high frequency domain.

The P_MF power as the marker of vegetative imbalance was demonstrated together with RR (cycle-length) and RR deviation trends. The "out" text file with parameters and the flight graph as the visual history are combined in one *png* file and archived by Paint software. At the different flight phases—independently from group categories—there were significant differences in implementing time periods (defining airspeed, approaching altitude, and attitude), so the time domain normalization was not achievable. Especially at landing tasks (repeated 10 times) the average heart rate is higher, than at take-off procedures—referring to increased vegetative sympathetic tone (Figs. 1 and 2). During focusing the attention the spectral power in middle frequency decreases, than slight rebound occurs: further oscillation is parallel with continuous vegetative tone modification.

The partition of en-route flight path is more difficult, so the exact identification of flight phases is not ready. (Specific integral software is necessary to calculate the area below the pulse and MF curves.) Retrospectively, the mental workload, the tenacity (focusing attention) and the low level of P_MF (rigidity) is noticeable; especially at approach and landing phase (Fig. 3).

Comparing to the resting phase the RR cycle change (the reciprocal changes of pulse) is similar in each group, that is the pulse increases, most explicitly during landing procedure (Fig. 4). The RR interval deviation at professional aviators increased, at the control students depressed (Fig. 5).

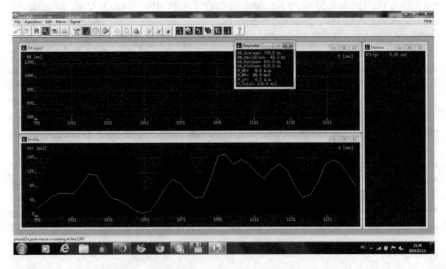

**Fig. 1** UAV operator 10 landing procedures (heart rate 78 bpm)

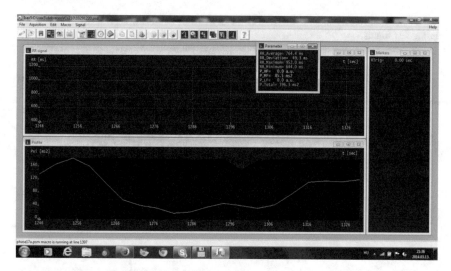

**Fig. 2** UAV operator 10 take off procedures (heart rate 80 bpm)

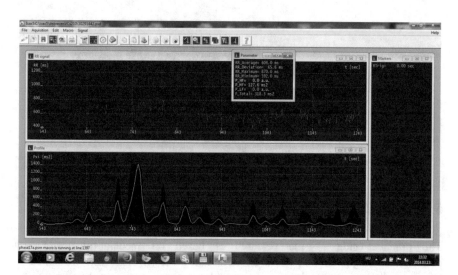

**Fig. 3** UAV Operator (82 % performance) en-route flight (12 min)

**Fig. 4** RR cycle changes during test flights

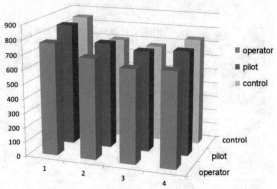

RR CYCLE CHANGES IN FUNCTION OF FLIGHT PHASES

1. Rest    2. Take offs    3. Landings    4. En-route

**Fig. 5** RR cycle deviation changes during flights

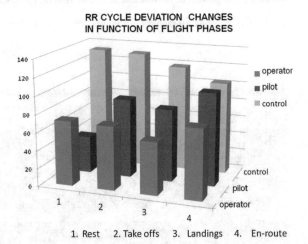

RR CYCLE DEVIATION CHANGES IN FUNCTION OF FLIGHT PHASES

1. Rest    2. Take offs    3. Landings    4. En-route

## 5  Physical Requirements for UAV Crews

The practical side of the research is based on medical-diagnostic and performance-diagnostic test methods. Conducting the physical tests we aimed to introduce the basic methodology of the analyses, the research results of measurements and the conclusions drawn from the statistical analysis. Our inducing factors for the physical research were:

– The importance to measure the level of development of skills, such as fine-motor coordination, sensory-motor coordination. Adequate physical reflections of special uses should be measured, which is one of the goals of the research.

- We have experienced that in the assessment of both training and performance conditional skills outweigh coordination skills.
- Emphasis should be shown on the development of complex coordination skills, such as fine-motor coordination, arm-leg coordination and the adaptability of the vestibular system.
- Measuring sensory-motor coordination; sense of balance, spatial and orientation skills; improving the capacity of the vestibular system; fine-motor coordination; reaction and movement time; coordination of the senses and the limbs.

Controlling UAV devices belongs to a moderate load zone based on physical activity (the type of muscle activity, energy consumption). During a UAV mission the human body is exposed to substantial nervous-mental load, which has significant influence on the efficiency of the activity. Senses (receptors) are heavily strained as well. Physical and nervous-mental load incurred by these activities predictably reduce ability to work. Professional work ability is likely to be closely related to the individual's physical state (physical preparedness). Therefore, developing and maintaining endurance is a primary part of the crew's physical training. As a result of current scientific research, other factors needed to improve the training will be apparent. A specific feature of the research allows us to measure the impacts of the conditional state when carrying out the activities. Thus, it makes the training of the crew more targeted and effective and aids in recruitment for this kind of job. According to one of our main tasks, performance structure has to be measureable from a professional, physiological and conditional perspective. In this phase of the research the conditional side is a priority. However, it can only be defined if the other components of the structure are known.

Research aimed to eliminate deficiencies in the experiment by:

1. Conducting a cross-sectional study is with a control group.
2. Using the same at each recording subjects.
3. In the final data processing only individuals with complete profiles were processed.
4. The conditions regarding the environment, the facilities, the equipment and their state were the same.

Teams involved in the research:
1. Examined group    50 pilots at Szolnok Helicopter Base, Hungarian Defense Force, students in the NFTC program
2. 12 UAV operator    Military/Soldiers
3. Control group    12 officer candidates, students at NUPS and secondary school students in Szolnok

**Table 1** Research test exercises [6]

| Dimension | Factor | Assigned test exercise |
|---|---|---|
| Cardiorespiratory endurance | Cardiorespiratory endurance | Cooper 12-min run test |
| Power | Static strength | Hand grip strength |
| | Explosive power | Standing long jump |
| Muscle strength | Functional arm strength | Bent-arm hang |
| | Trunk strength | Sit-up |
| Speed | Running speed | $10 \times 5$ m shuttle run test |
| | Speed of limb movement | Plate tapping |
| Joint mobility | Joint mobility | Sit-and-reach |
| Balance | Total body balance test | Flamingo test |

## 6   The Mathematical-Statistical Methods of Data Processing

The data analysis was conducted regarding the age, background, former flying experience, and general aviation knowledge of the participants. It is probable that the correlation among these features can lead to a conclusion of the best, the most capable, and trainable source of future UAV pilots.

Our analysis involved the followings:

- calculating the mean of the measured data
- calculating standard deviation of the average
- calculating standard deviation
- calculating the coefficient of variation
- correlation calculation
- calculating confidence interval, significance

Based on the analysis, the tests connected to the research topic have been com-piled, which can be seen in Table 1.

## 7   Findings on Aeromedical and Physical Tests

During the aeromedical and physical tests the research concluded on the following statements below:

- The stress level change in vegetative nervous system and pulse rate variability as an adaptive response is widely studied. This presented experimental setting is not for cardiologic purpose, only to reify the continuous changes in vegetative nervous system during stressful simulated military flight task. Due to the low case number we analyze only trends, our preliminary data are not sufficient to state any finite correlation with performance (flying success rate). The mental workload is accompanied by pulse rate increase at each group (as a tendency it was enhanced during landing procedure rather than take-offs). The pulse rate deviation increases at professional groups mirroring the psychic adaptation, while at students falls at complex tasks requiring constant attention.

- This research method belongs to the promising neuropsychology subject. Might be a useful tool to visualize the shifts in vegetative tone during mental workload. We should reach a full integration of time-frames (partition of pi-lot work into flight phases) with synchronized real time (quasi real time) data processing. After this standardization this model can contribute to the complex evaluation of operator's performance.
- The study of relationship between anthropometric features, the circulatory system, the conditional skills and complex skills gave results corresponding to the relevant literature.
- The significant relations between the measured parameters—particularly the conditional-coordination skills—suggest that relations between general health, the level of skills and performance grow stronger and deeper with age.
- Similarly to anthropometric indicators, indicators of the circular system show that the tested groups generated similar results to those who do not exercise regularly. The measured parameters show that 19 % of the research subjects have increased resting heart rate and high blood pressure.
- According to the Cooper test, the level of endurance is satisfactory, on the whole. However, in the evaluation of the test we have to bear it in mind that it compares performance to that of an average man. The expectations toward the tested groups are much higher than the average level of skills. From this perspective, their performance is rather poor.

# 8 Conclusion

UAV's application is no longer marginal within the aviation system, since recently, they are used in numerous fields and in various ways. The quick proliferation of remotely piloted vehicles raises many questions as to the methods of integration or flight safety before their integration. While the UAV itself is getting more and more sophisticated and reliable as a machine. The human factor, the operator, has received far less attention. The research discussed herein puts the human element into focus. Besides the theoretical and practical training issues, which have been studied but not displayed in this publication, we attempted to outline some core factors of UAV HF, for example, the physical or psychical suitability for the job. The data obtained from the simulator and tests will induce further studies on aeromedical or physical selection of future UAV crew members.

**Acknowledgements** This publication was created with the support of the European Union, and the co-financing European Social Fund under "Critical Infrastructure Protection" Research "TÁMOP-4.2.1.B-11/2/KMR-2011-0001".

# References

1. Farley, R., Heupel, K., Lee, K., Gardetto, P., Johnson, B.: Human factors in remotely piloted aircraft (RPA). HQ AFSC/SEHI DSN 246-0880, ASMA Annual Conference (Phoenix, Arizona) (2010). www.asma.org. Retrieved 06 Dec 2010
2. Lang, E., Horvath, G.: Integrated system for ambulatory cardio-respiratory data acquisition and spectral analysis (ISAX). User's Manual, Budapest, Hungary (1994)
3. Sayers, B.: Analysis of heart rate variability. Ergonomics **16**, 17–32 (1973)
4. Akselrod, S., et al.: Power spectrum analysis of heart rate fluctuation: a quantitative probe of beat-to-beat cardiovascular control. Science **213**, 220–223 (1981)
5. Mulder, G., Mulder-Hajonides Van Der Meulen, W.R.E.H.: Mental load and the measurement of heart rate variability. Ergonomics **16**, 69–83 (1973)
6. Izso, L.: Developing evaluation methodologies for human-computer interaction, Chap. 4, p. 88. Delft University Press, Delft, The Netherlands (2001)

# Simulation of Laser Alloying Process

**Imre Felde, Zoran Bergant and Janez Grum**

**Abstract**  The aim of the paper is to develop a simplified numerical model to predict the formation of the melt pool and the heat affected zone in single track laser alloying of C45 steel with NiCrBSi powder. The developed finite element model is based on the temperature field calculation using Fourier equations. The unknown coefficients such as surface absorption coefficient, volumetric efficiency and beam distribution coefficients are set according to cross-section geometry data, obtained from laser alloying experiment. The Nd:YAG solid state laser with multi-jet nozzle laser head with shielding gas argon was used to conduct experimental runs. The full factorial experimental design was used to evaluate the influence of power and scan feed rate on remelted cross-section area and microstructure. The calculated height and depth of melt pool and heat affected zone are in fairly good agreement with the experimental data. The presented numerical model require further refinement in order to take into account the complex physical phenomena during laser melting and alloying.

**Keywords**  Laser alloying · Remelting · Simulation · Steel · NiCrBSi

## 1  Introduction

Laser alloying is a process of melting a material with a high-power laser and then adding other alloying elements into the molten pool or with pre-deposited layer [1]. Surface alloying and cladding when performed by a laser, require minimal post-process re-machining of the surface and localized residual stresses [2]. The

I. Felde (✉)
Faculty of Informatics, Óbuda University, Budapest, Hungary
e-mail: felde.imre@nik.uni-obuda.hu

Z. Bergant · J. Grum
Faculty of Mechanical Engineering, University of Ljubljana, Ljubljana, Slovenia
e-mail: zoran.bergant@fs.uni-lj.si

J. Grum
e-mail: janez.grum@fs.uni-lj.si

© Springer International Publishing Switzerland 2016                                                     59
L. Nádai and J. Padányi (eds.), *Critical Infrastructure Protection Research*,
Topics in Intelligent Engineering and Informatics 12,
DOI 10.1007/978-3-319-28091-2_5

laser alloying operating window is defined in terms of laser beam mode, power, spot diameter, scanning speed and powder feed rate. Only values of parameter setting within limited ranges can be applied to generate tracks meeting the geometrical requirements to produce crack and pore free coatings by overlapping single tracks [3]. The thermo-kinetic laser powder deposition model coupling finite element heat transfer calculations with transformation kinetics and quantitative property–microstructure relationships was developed [4].

Surface alloying with a laser is similar to laser surface melting with the exception that another material is incorporated into the melt pool. In most modeling applications, assumptions must be made to reduce the complexity of the physical phenomena and to reduce the amount of pre- and post-processing time. The simplified temperature model for prediction of area cross section in laser alloying where the heat input is presented using the idea of Guo and Kar [5]. The temperature field was calculated using the Fourier equation and under assumption that the molten pool is half-ellipsoid. The numerical values were set according to experimental data to adjust the variables such as absorption coefficient, volumetric efficiency and the beam distribution coefficient.

## 2  Temperature Field Calculation

The numerical model is based on the calculation of the temperature in the substrate during the laser treatments. As a first approximation, the melt pool region as well as the heat affected zone is assumed to be in agreement with the peak temperatures obtained by the laser beam in the whole volume of the specimen. The temperature field was calculated by solving the Fourier formula,

$$\rho C_p \frac{\partial T}{\partial t} + S \rho C_p \frac{\partial T}{\partial r} = \frac{\partial}{\partial r}\left[k(T)\frac{\partial T}{\partial r}\right] + Q \tag{1}$$

where r vector denoted the local coordinates, $T(r, t)$ is the temperature at any point of the substrate at a given time t to the velocity S of the laser beam moving above the fixed specimen, $\rho$ is the density, $C_p$ is the specific heat, k is the thermal conductivity and Q is the internal heat generation per unit time and unit volume. Equation (1) should be satisfied using the initial and boundary conditions. At time $t = 0$ the substrate has a uniform temperature ($T_0$) throughout its volume

$$T(r, 0) = T_0 \tag{2}$$

The boundary conditions are:

$$-k(T)\left(\frac{\partial T}{\partial r}\right) = h[T - T_{am}] + e\sigma[T^4 - T_{am}^4] - q_s \tag{3}$$

where h and e is surface heat transfer coefficient and emissivity respectively; $\sigma$ is Stefan–Boltzmann constant, and $T_0$ is the ambient temperature. The term $q_s$ stands for the imposed heat flux onto the surface due to the laser beam. The heat input into the substrate due to the application of laser beam is considered in terms of surface heat flux [5] through the term $q_s$ in equation until the top surface of substrate is below solidus temperature as

$$q_s = \frac{P\eta_{gauss}d}{\pi R_{eff}^2} \exp\left(-\frac{dR^2}{R_{eff}^2}\right) \tag{4}$$

where P refers to beam power, $\eta_{Gauss}$ absorption coefficient, $R_{eff}$ the effective radius of the laser beam and d is the distribution coefficient related to the pattern of the laser beam profile. As the substrate subsequently melts, the surface heat flux is replaced by a volumetric heat source expression for the portion of the molten substrate as

$$\dot{Q} = \frac{6\sqrt{3}P\eta_{vol}}{\pi\sqrt{\pi}a^2 b} \exp\left(-\frac{3R^2}{a^2} - \frac{3z^2}{b^2}\right)$$

where

$$
\begin{aligned}
a &= R_{eff} \quad for \quad a \leq R_{eff} \\
a &= w_i \quad for \quad a > R_{eff} \\
b &= p_i
\end{aligned}
\tag{5}
$$

where $w_i$ and $p_i$ are instantaneous values of the computed width and penetration of the liquid pool respectively performed by the numerical calculations. The absorption coefficient of volumetric heat is $\eta_{vol}$. The shape of the volumetric heat source was assumed to an ellipsoid having axis a and b. The heat input taken into account from Eq. (5) is applied directly into the melt pool volume through the term in Eq. (1).

## 3   Materials and Methods

Experiments were carried out on substrate made from C45 steel (Mat. No. 1.0503). Figure 1a shows the network of ferrite-pearlite microstructure of substrate. Figure 1b shows nickel-based powder NiCrBSi, which was deposited in the molten pool during laser alloying through the nozzle of the laser head. Chemical compositions of the base and feedstock material are given in Table 1.

The specimens were laser alloyed at Bay Zoltán Institute in Budapest using the solid state Nd:YAG laser (Rofin), powder feeder system with carrier gas (Argon), laser cladding head with circumferential arrangement of 4 nozzles injection. The

(a)                                                    (b)

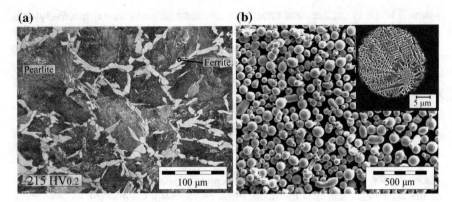

**Fig. 1** Base and feedstock material, **a** C45—0, 45 % C steel microstructure, **b** NiCrBSi gas atomized power

**Table 1** Chemical compositions of the substrate and the powder (weight %)

|                  | C    | Cr   | Ni   | Si   | Mo  | Mn   | B    |
|------------------|------|------|------|------|-----|------|------|
| C45 steel 1.0503 | 0.46 | 0.40 | 0.40 | 0.40 | 0.1 | 0.65 | –    |
| NiCrBSi powder   | 0.51 | 13   | 78.1 | 3.22 | –   | –    | 2.08 |

laser track length was L = 50 mm and the cross-section was examined on the plane with offset of L/2 = 25 mm from the laser start. The power levels were 1000, 1500, 2000 and 2500 W and scanning speeds were 400, 500, 600 and 700 mm/min.

Metallographic specimens were prepared following standard procedure and etched in nital solution. The cross-section images were acquired using optical microscope. The chemical analysis was carried out with JEOL JXA 8600 scanning electron microscope using energy dispersive spectroscopy.

## 4  Results and Discussion

The temperature model has been discretized in the COMSOL finite element model code. The real geometry of the substrate has been applied in the FEM model where the laser is scanned in x direction. Values of absorption coefficient, $\eta_{Gauss}$, volumetric efficiency $\eta_{vol}$ and beam distribution coefficient are considered to be 0.30, 0.35 and 3.0 to compute melt pool dimensions corresponding to all the experimental laser alloying conditions. In the model we assumed the melting of the substrate occurs when the maximum temperature on the surface reaches liquidus point.

The calculated temperature distributions at 3.25 s after the beginning of laser irradiation, are given in Figs. 2 and 3. High, non-equilibrium heating rates suggests the transformation points $A_{c1}$ and $A_{c3}$ shift to higher temperature level. The cooling time $t_{8/5}$ is in a range from 0.5–0.8 s which is a possible reason for generation of cracks in the melted zone (Fig. 4b).

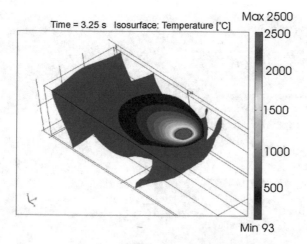

**Fig. 2** The calculated temperature distribution in the work piece

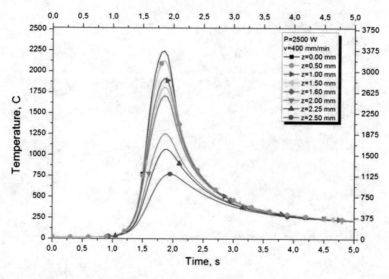

**Fig. 3** The calculated temperature as functions of time and distance measured from the surface at 10 mm from the starting location of the laser process

The geometry of cross-section was digitalized and the basic dimensions were taken into account, area cross-section, depth, width of melted and heat affected zone. The relationship of the remelted area A (mm$^2$), power (P) and scanning speed (S) is shown in Fig. 5. The remelted area A (mm$^2$) is increasing with power and decreasing with scanning speed S. Figure 5 shows the macro and microstructure of alloyed region.

The temperature transformation austenite (TTA) diagrams are used to predict the formation of austenite in the heated zone. The heating rate with laser is in a range of

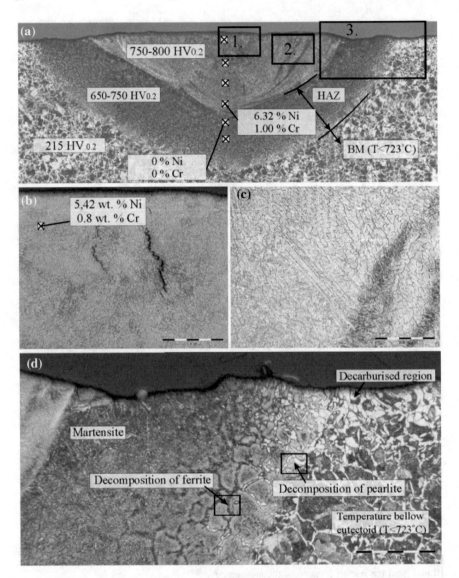

**Fig. 4** Microstructure of alloyed zone and substrate. **a** Macrograph of cross-section, specimen P = 1000 W, v = 600 mm/min. **b** Cracks in melted and alloyed zone. **c** Melted and alloyed zone. **d** Transition microstructures in the heat affected zone

1000 °C/s, therefore the formation of austenite is at higher temperature than with very slow heating rate (in furnace). The temperature where the austenite starts depends on the heating rate. When the rate of heating increases, the austenite formation takes place at the higher temperature. The TTA diagram for C45 steel is presented in Fig. 6a. The energy dispersive spectroscopy linescan was made through each cross-section of metallographic specimen. The linescan values were

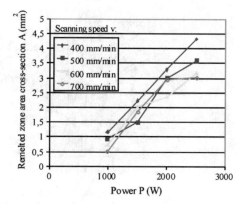

**Fig. 5** The remelted area as a function of laser power

averaged of each sample. The correlation of chemical composition and laser scanning parameters were not found when comparing the individual tracks made with different parameters. The average values, used for further analysis of alloyed material show the average composition of 0.45 % C, 5.4 % Ni, 1.07 % Mn, 0.96 % Cr, 0.62 % Si., 0.1 % Mo. Figure 7 shows the TTA diagram for alloyed material. From the comparison of each diagram it is evident that the austenite on heating forms at lower temperature in the alloyed region than in C45 steel.

On macrograph of Fig. 4, the decomposition of pearlite is visible on the elliptical $A_{c1}$ isotherm. We assume this temperature corresponds to around 830 °C according to the TTA diagram for heating rate of 1000 °C/s, Fig. 6b. Moreover, isothermal line, where ferrite grains disappear indicate the $A_{c3}$ isotherm which is at around 930 °C. The melted and alloyed region was subsequently subjected to high cooling rate therefore martensite was formed over the entire area. The 100 % martensite also formed near the fusion line in the heat affected zone. High cooling rate and the metallurgical effect of alloying contributed to hardness between 750 and 800 HV.

Figure 8 shows the comparison of remelted area cross-section of the model and experimental data. The lack of fit between the molten pool was observed at 1 and

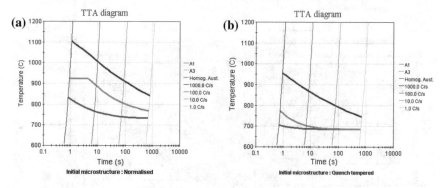

**Fig. 6** Temperature transformation austenite (TTA) diagram for C45 composition (**a**) and for alloyed region averaged composition (**b**) (generated by JMatPro)

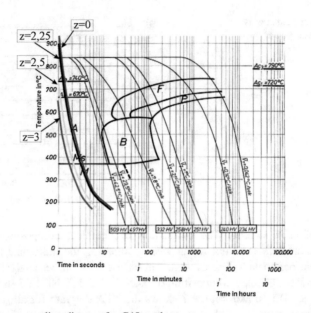

**Fig. 7** Continuous cooling diagram for C45 steel

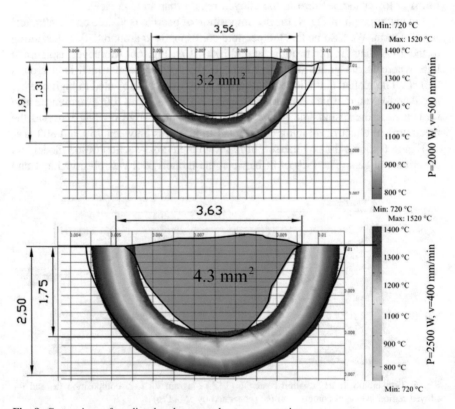

**Fig. 8** Comparison of predicted and measured area-cross section

1.5 kW. On the other hand a good agreement between measured and predicted values is in the range from 2 to 2.5 kW. From the comparison it is clear that the presented numerical models require further refinement to order to take into account the complex physical phenomena during laser melting and alloying.

# 5 Summary

The surface properties after laser alloying of C45 steel with NiCrBSi self-fluxing gas atomized powder were studied and the simplified temperature-time model was proposed. The real geometry of the substrate has been applied in the finite element model. The microstructure after melting and alloying is brittle and susceptible to hot cracking.

The predicted peak-temperature field is, in most cases, in a good agreement with the microstructural variation from cross-section images. However, in some cases the width and the depth of the melted pool determined at real samples differ from the calculated ones. The reason of the discrepancy could be originated to the simplicity of the model which does not include several phenomena occur during the laser alloying of the substrate.

**Acknowledgements** We acknowledge the financial support of this work by the Hungarian State and the European Union under the TÁMOP-4.2.1. B -11/2/KMR-2011-0001 project.

# References

1. Toyserkani, E., Khajepour, A., Corbin, S.: Laser Cladding, Taylor and Francis. CRC Press, UK (2005)
2. Grum, J., Žnidaršič, M.: Residual stress analysis after laser surface alloying with various powdered materials. Int. J. Microstruct. Mater. Prop. **1**, 219–230 (2006)
3. Felde, I., Réti, T., Kalazi, Z., Costa, L., Colaço, R., Vilar, R., Verő, B.: A simple technique to estimate the processing window for laser clad coatings. In: 1st ASM International Surfacing Engineering Conference and the 13th IFHTSE Congress, pp. 237–242 (2002)
4. Costa, L., Vilar, R., Reti, T., Colaço, R., Deus, A.M., Felde, I.: Simulation of phase transformations in steel parts produced By laser powder deposition. Mater. Sci. Forum **473–474**, 315–320 (2005)
5. Guo, W., Kar, A.: Determination of weld pool shape and temperature distribution by solving three dimensional phase change heat conduction. Sci. Technol. Weld. Joining **5**, 317–323 (2000)

# Hybrid Optimization Approach for Determination of Thermal Boundary Conditions

Imre Felde

**Abstract** The estimation of thermal boundary conditions occurring during heat treatment processes is an essential requirement for the characterization of heat transfer phenomena. In this work, the performance of five optimization techniques is studied. These models are the Conjugate Gradient Method, the Levenberg-Marquardt Method, the Simplex method, the NSGA II algorithm and a hybrid approach based on the NSGA II and Levenberg-Marquardt Method sequence. The models are used to estimate the heat transfer coefficient in 2D axis symmetrical case during transient heat transfer. The performance of the optimization methods is demonstrated using numerical experiments.

**Keywords** Quenching · Inverse heat conduction problem · Hybrid optimization

## 1 Introduction

Immersion quenching is widely applied in industry to change a materials properties under high temperatures and high rates of cooling, a condition in which the heat transfer can be dominated by the cooling characteristics o the cooling media. To attain the required heat transfer conditions, exactly known thermal loads must be supplied on the material surface, namely design surface, and thus both the temperature and the heat flux are prescribed. The problem consists of determining the specifications for the heat exchange to achieve the desired conditions on the design

I. Felde (✉)
Faculty of Informatics, Óbuda University, Budapest, Hungary
e-mail: felde.imre@nik.uni-obuda.hu

© Springer International Publishing Switzerland 2016　　　　　　　　　69
L. Nádai and J. Padányi (eds.), *Critical Infrastructure Protection Research*,
Topics in Intelligent Engineering and Informatics 12,
DOI 10.1007/978-3-319-28091-2_6

surface. Performing inverse heat conduction problem (IHCP) analysis [1–4] the required information for the heat transfer process can be achieved.

The IHCP of immersion quenching has been usually tackled by either implicit or explicit formulation. In the implicit approach, the problem is formulated as a multivariable optimization problem, while the explicit formulation attempts to determine directly the unknown parameters with the use of regularization techniques to solve the resulting system of equations. There are two distinct groups of optimization techniques: the deterministic methods (Conjugate Gradient [3], Levenberg-Marquardt [4], Simplex [5], etc.) and the stochastic approaches (genetic algorithms [6], particle swarm optimization [7, 8] etc.).

In general, deterministic methods are faster than stochastic methods, although they are more prone to converge to a local instead of the global minima or maxima. On the other hand, stochastic algorithms, despite being more likely to converge to the global minima or maxima, are in general, expensive computationally. Various optimization techniques have been applied to estimate the heat transfer during quenching process as well.

This study focuses on a hybrid solution that combines two approaches: a stochastic method, by which the global extremum in the search space can be localized and the deterministic formulation, which is for the swift find, the global optimum.

## 2   The Heat Conduction Model

The mathematical formulation of the transient heat transfer for a homogeneous isotropic domain ($\Omega$) is defined as follows:

$$\nabla \cdot (k(\mathbf{r}, T) \cdot \nabla T) + Q(T, \mathbf{r}, t) = C_p(\mathbf{r}, T)\rho(\mathbf{r}, t)\frac{\partial T}{\partial t} \tag{1}$$

where $\mathbf{r} \in \Omega$ is the spatial vector, t is the time, k is the thermal conductivity, T is the temperature, $C_p$ is the specific heat, $\rho$ is the density and Q is the latent heat. The initial condition is

$$T(\mathbf{r}, t = 0) = T_0(\mathbf{r}) \tag{2}$$

where $T_0$ is the initial temperature of the domain. The boundary conditions are expressed by:

$$-k\frac{\partial T}{\partial \mathbf{r}} = HTC_i(T(\mathbf{r}, t) - T_{am}) \text{ in } \Gamma i \; i = 1\ldots p \tag{3}$$

where HTC$_i$ $hi$ are the heat transfer coefficients corresponding to different portions of the boundary ($\Gamma_1 \cup \Gamma_2 \ldots \cup \Gamma_p = \Gamma$ and $\Gamma_1 \cap \Gamma_2 \ldots \cap \Gamma_p = \emptyset$) and $T_{am}$ is the ambient temperature.

# 3 The Inverse Heat Conduction Model

Assuming that the temperature inside the work piece and/or on its surface is measured during the heat transfer process, it is possible to solve the inverse heat conduction problem by determining the time/or temperature variations of the thermal boundary conditions [1–3]. Each one of domain boundary zones $\Gamma$, is considered to have a time dependent heat transfer coefficient, HTC(t). The time dependence of the heat transfer coefficient can be approximated by polynomial functions, each one defined by a set of parameters $h_i^{(r)}$ = (r = 1…p; i = 1…q), according to Fig. 1. The unknown design parameters can be expressed by the vector of $m$ ($m$ = $p$ * $q$), components $\tau = (\tau_1, \ldots, \tau_m) = \left( h_1^{(1)}, \ldots, h_q^{(1)}, h_1^{(2)}, \ldots, h_q^{(2)}, \ldots, h_1^{(p)}, \ldots, h_q^{(p)} \right)$. The temperature at different times is given by measurements at $n$ points in the solid region, located at $r_k$, (k = 1…n). On calling $T_k^m$, the measured temperatures, and $T_k^c$, the calculated temperature at those points, one can pose the problem of obtaining the values of the heat transfer coefficients $\tau_i$ that minimize the cost function, S:

$$S = S(\tau_1, \ldots, \tau_m) = \sum_{k=1}^{n} \left( T_k^m - T_k^c \right)^2 = \min \qquad (4)$$

where $n$ is the total number of measured temperatures, i.e., the number of points multiplied the number of measurements at each point.

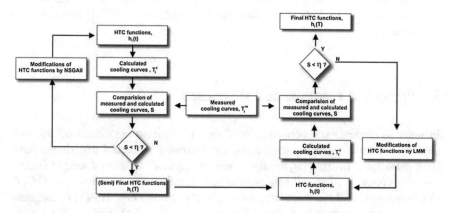

**Fig. 1** The iterative procedure for the determination of thermal boundary conditions

## 4   Hybrid Formulation

The solution of the inverse parameter estimation problems is based on the minimization of Eq. (4). The following optimization approaches have applied used to minimize the value of S:

1. Simplex (**Simplex**) search method is based on the Nelder–Mead algorithm [5]
2. Levenberg-Marquardt Method (**LMM**) [4]
3. Conjugate Gradient Method (**CGM**) [3, 4, 9]
4. The Non-dominated Sorting Genetic Algorithm (**NSGA II**) [10–12]
5. Hybrid method using (**NSGA II**) and (**LMM**) is sequentially

The following steps are obtained by using the optimization methods if they used by their own (Simplex, LMM, CGM and NSGAII) according to Fig. 1:

1. During the initial iteration, the components of the vector $\tau$ are initialized to some values
2. The values of $HTC_i(t)$ functions are set
3. The cooling curves $(T_k^c)$ are calculated based on numerical simulation
4. The difference between the measured $(T_k^m)$ and the calculated $(T_k^c)$ time-temperature signals are characterized by calculating S
5. If the value of S is greater than a desired tolerance value $(\eta)$ then the $h_i(t)$ functions are modified by using the optimization algorithm and a new iteration is started at Step 2. If S is less than $\eta$ then the iteration stops.
   The Hybrid approach is also based on the Steps 1–5, while the output results of Step 5 will be the input data for the temperature field computations. The Hybrid method requires the following additional steps:
6. The cooling curves $(T_k^c)$ are calculated based on numerical simulation
7. The difference between the measured $(T_k^m)$ and the calculated $(T_k^c)$ time-temperature signals are characterized by calculating S
8. If the value of S is greater than a desired tolerance value $(\eta)$ then the $h_i(t)$ functions are modified by using the LMM algorithm and a new iteration is started at Step 6. If it is so, then the final estimated $HTC_i(t)$ functions are estimated

## 5   Numerical Example and Discussion

In order to compare the performance of the optimization algorithms on the prediction of thermal boundary conditions, the following numerical experiment have been performed. In the analysis, there was no physical set-up to directly measure the temperature $T_k^m$. Instead, a theoretical heat transfer coefficient function $HTC(T, z)$ was assumed and has been substituted directly into the Eqs. (1)–(3) to calculate the temperatures at each location for the thermocouples (TCs). The results are used in the computed temperature $T_k^m$ curves. Due to this concept the $T_k^m$ curves have been assumed to be error-free samples. The following concept have been used for the computational investigations:

- The theoretical HTC(T, z) function have been determined
- The $T_k^m$ temperature signals have been generated by obtaining simulations on the basis of HTC(T, z) functions
- Inverse computations have been carried out by applying each optimization method, in order to reconstruct the original HTC(T, z) function
- The computational results were analyzed

The quenching process for a cylindrical work piece, mounted with 5 TC's was investigated. A 2D axis-symmetric heat transfer model was applied to calculate the temperature distribution during the cooling process. The physical properties of the Iconel 600 alloy were assigned to the work piece [13]. The thermocouples were assumed to be mounted at 1 mm below the side surface of the rod. The locations of the TC's were assumed at 1 mm below the surface at 0, 50, 150, 200 and 250 mm distance measured form the bottom of the cylinder respectively to TC 1-5. The parameters used for the calculations are summarized in Table 1.

The effect of wetting front kinematics occurs during immersion quenching, is taken into consideration by defining the heat transfer coefficient [14, 15], Eq (5). HTC(T, z) used is functions of temperature and the vertical local coordinate. The theoretical heat transfer coefficient function is represented at Fig. 2., while the cooling curves obtained at the TC locations are shown at Fig. 3. The HTC(T, z) is used for all the surfaces of the work piece including the top and the bottom faces as

| Table 1 Parameters applied for the computational example | | |
|---|---|
| Radius, R | 25 mm |
| Length, L | 200 mm |
| Initial temperature, $T_0$ | 850 °C |
| Ambient temperature, $T_{am}$ | 50 °C |

**Fig. 2** The theoretical heat transfer coefficient function applied

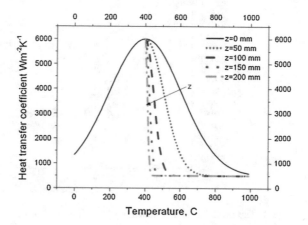

**Fig. 3** The predicted cooling
curves obtained at the TC
locations in direction of axis
"z"

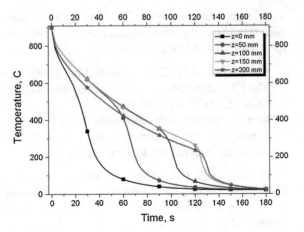

well. For the inverse calculations 100 components of the vector τ have been
applied, while the initial estimate of HTC(T,z) was set to 100 Wm$^{-2}$ K$^{-1}$. The
population size was defined as 100, for the NSGA II method.

$$h_i(T, z) = \begin{cases} 5500 * e^{-e^{-11.34*(T-400)^2}} + 500 & T \leq 400\,°C \\ 5500 * e^{-e^{(-11.34 + 0.0606*z - 2.653*10^{-4}*z^2)*(T-400)^2}} + 500 & T > 400\,°C \end{cases} \quad (5)$$

In order to reconstruct the thermal boundary conditions 500 generations were
investigated by the NSGA II model and 400 iterations were performed by the
LMM, CGM and Simplex models (Fig. 4). The concept of using the Hybrid
approach was to perform the generations until the cost function exceeds 2.0e4, or

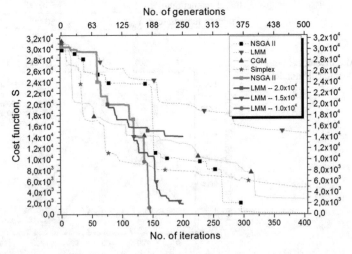

**Fig. 4** The evolution of cost function (S), as functions of iterations and generations

1.5e4, 1.0e4 and then the iteration was continued by means of LMM. Then the LMM calculations were carried out till the S was lower than 100 or the number of LMM iterations exceed 150.

The results obtained by Simplex, LMM and CGM showed a limited accuracy for the inverse estimation. The discrepancy of the predicted heat transfer coefficient were still unacceptable huge after the 400 iterations. Very good agreement between

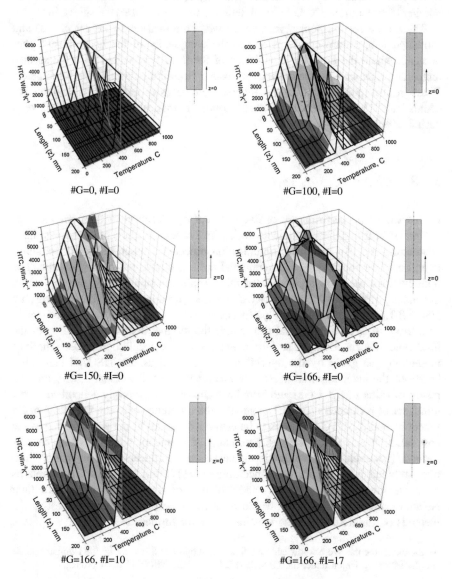

**Fig. 5** The original and reconstructed HTC(T, z) are presented after a certain number of generation (#G) and iteration (#I)

the original and recovered HTC functions was given by using the NSGAII algorithm. However, at least 300 generations had to be constructed to achieve the desired output. The fastest convergence with the highest recovery performance was given by using the hybrid optimization sequence when the NSGAII method was used until the value of S was lower than 1.0e4. Similar results but slower convergence obtained in the case of LMM—1.5e4, while rather big S belongs to LMM—2.0e4. The estimated thermal boundary functions obtained by using the hybrid approach at given number of generations (NSGAII) and iteration numbers (LMM) are shown in Fig. 5.

The reason for the poor agreement between the measured (pre-calculated) and estimated temperature curves is due to the complexity of the heat transfer phenomena, where the boundary conditions of the third type varied with the surface temperature, as well as, the distance measured from the bottom of the work piece. These results point out that the inverse heat transfer calculations applied for sophisticated thermal problems needs robust numerical methods to achieve a desirable outcome.

# 6  Summary

In this work, the performance of five different optimization models, for the estimation of heat transfer coefficients during an immersion quenching process, have been compared. An automatic optimization procedure, based on a process simulator, cost function and various numerical optimization techniques was used. The optimizations methods applied were the Simplex, Conjugate Gradient Method, Levenberg-Marquardt Method, NSGAII method and a hybrid approach based on a NSGAII-LMM sequence. The performance of the optimization algorithms is compared using on a numerical test, where the thermal boundary condition of the third type was functions of surface temperatures and local coordinates. The best prediction was given by the hybrid method, as well as, the NSGAII algorithm, however, the latter required more computational effort. It must be noted, that the results of using LMM, CGM and Simplex techniques are strongly dependent on the initial set of parameters [6, 7]. The smaller the difference between the initial guess functions and the HTC functions to be estimated, the faster the convergence of the cost function and more accurate prediction of boundary conditions can be performed. However, this investigation is based on the concept, that no any preliminary knowledge has been given concerning the HTC functions. These assumptions comprise the parameter intervals and isolation of the search space. With finer isolations, the applied methods would most likely perform differently. Testing the methods on different isolations remains a task for further investigation.

**Acknowledgements** We acknowledge the financial support of this work by the Hungarian State and the European Union under the TÁMOP-4.2.1. B -11/2/KMR-2011-0001 project.

# References

1. Beck, J.V., Blackwell, B., St Clair, C.R., Jr.: Inverse Heat Conduction. Wiley, New York (1985)
2. Tikhonov, A.N., Arsenin, V.Y.: Solution of Ill-posed Problems. Winston, Washington, DC (1977)
3. Alifanov, O.M.: Inverse Heat Transfer Problems. Springer, Berlin/Heidelberg (1994)
4. Özisik, M.N., Orlande, H.R.B.: Inverse Heat Transfer: Fundamentals and Applications. Taylor & Francis, New York (2000)
5. Nelder, J.A., Mead, R.: A simplex method for function minimization. Comput. J. 7(4), 308–313 (1965)
6. Lagarias, J.C., Reeds, J.A., Wright, M.H., Wright, P.E.: Convergence properties of the Nelder-Mead simplex method in low dimensions. SIAM J. Optimiz. 9(1), 112–147 (1996)
7. Das, R.: A simplex search method for a conductive–convective fin with variable conductivity. Int. J. Heat Mass Transf. 54, 5001–5009 (2011)
8. Nelles, O.: Nonlinear System Identification. Springer, Berlin (2001)
9. Fletcher, R., Reeves, C.M.: Function minimization by conjugate gradients. Computer J. 7, 149–154 (1964)
10. Deb, K., Pratap, A., Agarwal, S., Meyarivan, T.: A fast and elitist multiobjective genetic algorithm: NSGA-II. IEEE Trans. Evol. Comput. 6(2), 182–197 (2002)
11. Deb, K.: Multi-objective Optimization Using Evolutionary Algorithms. Wiley, Chichester, UK (2001)
12. Veldhuizen, D.A.V., Lamont, G.B.: Multi-objective evolutionary algorithms: analyzing the state-of-the-art. Evol. Comput. 8(2), 125–147 (2000)
13. Clark, J., Tye, R.: Thermophysical properties reference data for some key engineering alloys. High Temp. High Pressures 35/36, 1–14 (2003/2004)
14. Majorek, A., Scholtes, B., Müller, H., Macherauch, E.: Influence of heat transfer on development of residual stresses in quenched steel cylinders. Steel research No. 4, pp. 146–151 (1994)
15. Tensi, H.M., Stick, A.: Martens hardening of steel—prediction of temperature distribution and surface hardness. Mater. Sci. Forum 102–104, 741–775 (1992)

# Security Research in the Field of Climate Change

József Padányi and László Földi

**Abstract** The paper aims at giving an overall picture about the effects of climate change on the military. It describes the results of a number of researches related to these challenges the military has to face. It argues that for military bases, the use of renewable energy (especially solar cells) is cost-effective, and in line with the modernization of the Army, 'Green barracks' programme has been launched in Hungary. The paper also reveals the importance of adequate military clothing for soldiers that may tackle the effects of the extreme ambient temperatures.

## 1 Preface

The work that is introduced herein is focused on areas of research that are significantly published abroad, but have only a few publications in Hungary. Although climate change is a "popular" topic, so far only a few experts have studied its effect on security and on application in the military. A publication titled "Climate change and military power" needs to be mentioned here, published in 2010 under the wing of the Strategic Defense Institute (Authors: László Kohut, József Koller, Gábor Lévay, József Padányi). This way of thinking was continued when our two-year research project was started.

The project team consisted of 3 Doctors of the Hungarian Academy of Sciences, 6 regular PhD's, 2 candidates for doctoral degree, 5 university students and 15 others; altogether 31 persons participated in the work. Our results were presented at several conferences and in scientific papers both in Hungary and abroad. In our opinion, it was an outstanding result that—in cooperation with the staff of the Sanitary Engineering Department of Pollack Mihály Technical and Informatics

J. Padányi (✉) · L. Földi
National University of Public Service, Budapest, Hungary
e-mail: padanyi.jozsef@uni-nke.hu

L. Földi
e-mail: foldi.laszlo@uni-nke.hu

© Springer International Publishing Switzerland 2016
L. Nádai and J. Padányi (eds.), *Critical Infrastructure Protection Research*,
Topics in Intelligent Engineering and Informatics 12,
DOI 10.1007/978-3-319-28091-2_7

Faculty, we completed the instrumental examination of all the articles of clothing that have been brought into service by the Hungarian Army.

Our most important purposes was to synthesize the known effects of climate changes, systematize the open issues, and adapt the results that have already proven to be effective to Hungarian conditions by exploring the results available in the domestic and international literature. Our definite goal was to increase the security of our soldiers in all areas of military life, from mission assignments and defense against disasters to the challenges of the day-to-day life.

In order to achieve these goals, we conducted research in the following areas:

- Processing the domestic and international literature, and organizing the knowledge in a data repository
- Facts of climate change, theories and scenarios
- Experience gained in applying military force in the area of defense against disasters
- Challenges in NBC defense represented by climate changes
- Changes in the energy supply of the Army
- Technical assets and climate change
- Healthcare aspects of global climate changes
- Effects of extreme climatic factors prevailing in operational areas on the psychic— mental performance of soldiers
- Using renewable energy sources in the Army
- Security and environmental issues related to the depletion of fossil energy sources
- Effect of transport on the environment. Opportunities for decreasing motor vehicle emissions in the Hungarian Army
- "Green barracks" program
- Opportunities for renewing facilities managed by the Ministry of Defense (Opportunities of applying methods of green architecture, and alternative energy sources)
- Effect of air, precipitation and soil related changes on security
- Secondary effects of the global climate change on the security of Carpathian basin
- Effect of the global climate change on the defense of critical infrastructure facilities
- Effect of the global climate change on agriculture
- Effect of climate changes, warming and extreme weather conditions on the application of technical assets and on their operation, repair, maintenance and life cycle
- Effects of extreme ambient temperatures on military clothing; directions of selection and development of military clothing

We would like to summarize some of our most important conclusions in the followings.

## 2 The "Green Barracks Program"

The "Green Barracks" program is a good example for the ongoing research and development within the Hungarian Armed Forces [1].

The building energy programs executed and planned in the defense forces and the ambitions to utilize renewable energy all serve the purpose of reducing both energy consumption and $CO_2$ emission. The modernization of heating systems, producing domestic hot water using solar energy and photovoltaic energy generation, installation of advanced building energetic systems, switch over to energy efficient, up-to-date lighting systems, the application of building management systems, the utilization of rainwater, etc. together make up this program.

From the point of energetic issues out of the military infrastructure properties with superstructures (barracks, institutions, warehouse bases, etc.) play an obvious role. The scale of the property pool managed by MoD means approximately 1700 facilities with 14,500 buildings and with a value of HUF 360 billion. The price of its annual energy consumption is more than HUF 8 billion, so there is a chance to save money.

Part of the program is the introduction of alternative energy sources (biomass, biogas, solar, wind and geothermal energies) to the supply of military facilities, use of renewable energy sources to generate electricity, hot water and heat. The use of green energies expands to the field of military operations and camps, because economic and security aspects of energy-supply have significant importance there. Many elements of military infrastructure are also elements of the national critical infrastructure, so any improvement of their security and energy supply serves in the protection of the critical infrastructures at the same time.

The green barracks program makes efforts to counterbalance energy consumption, which is increasing in connection with the climate change, in the area of improving and maintaining military infrastructure and at the same time to enforce the utilization of renewable energy and the environmental aspects.

## 3 Effect of Extreme Weather to Psychic-Mental Performance

Proper handling of environmental factors is the essence of military operational success and survival. Detailed knowledge of special mechanisms is extremely important, as heat affects the cognitive, behavioral, and subjective reactions of soldiers. Heat stress significantly reduces military performance and psychological changes can predict critical physiological symptoms. During our research we investigated the mental, psycho physiological, physiological and group-psychological elements of psychological performance under a hot climatic environment. We investigated the changes of cognitive performance, reaction time, sense perception, watchfulness, complex mental and psycho-motor performance, dexterity, stamina, targeting, target tracking, simultaneous tasks, subjective reactions, sensible symptoms and phenomena, sleeping cycle, group-psychological processes and emotional readiness under high environmental temperatures [2].

Our key conclusions are contained in the following:

- Military leaders must be provided basic physiological and psychological training in connection with the effects of extreme climates on military performance to be able to perform the operational planning of missions and the preparation of the military personnel.
- Commanders working in mission areas must be provided relevant data regarding the local climate and the expected physiological and psychic effects thereof for the planning of each mission and the regular service activities.
- During missions performed under extreme temperature conditions that are longer than 3–6 h, psychic performance is expected to decrease. This impact can be mitigated by dividing complex tasks into sub-tasks and by simplifying communication.
- The commanders controlling the operations must give special care to their staff in case of activities performed in extreme temperatures, in wet environments or in water, because long-term exposure may lead to more rapid and significant performance reductions and injuries.
- It is recommended for commanders of missions to include in the practice activities in the territories with a given climate tasks providing information on the current status of psychological performance (with special regard to the use of special protective clothing and equipment). They have to check whether the clothes and protective equipment worn and the equipment, devices meet the requirements of the given climate in order to protect and maintain the health of the personnel.
- When designing and producing military and protective clothes, extreme climatic factors, as well as scientific and practical experience gained from international research must be taken into account (along with the experience of soldiers who served in missions), with special emphasis on the characteristics of the micro-environment forming between the body and the clothes worn and its role in thermoregulation.
- As far as special protective clothing is concerned, the personnel must be provided with equipment that already incorporates state-of-the-art technological solutions managing the negative impacts of perspiration and humidity and providing proper thermal insulation.
- The sanitary corps must be prepared to detect and treat the symptoms and phenomena of both thermal stress and hypothermia. In the mission areas, they have to maintain the basic knowledge of soldiers within the framework of self-help and companionship.
- Under extreme temperature conditions, the manual skills of grasping and holding deteriorate, and it becomes more difficult to control certain devices and equipment manually due to the overheating or overcooling of controls. This justifies the redesign of some military devices and equipment in accordance with

the peculiarities of the given climate, also taking into consideration ergonomic aspects.

# 4 Effects of Climate Change to CBRN Defense

Climate change affects people, installations and technical equipment. Climate change has an impact on the CBRN defense a little bit more significantly than other services in military forces. Due to predictably increasing occurrences of extreme weather conditions, domino effect of the natural meteorological and hydrological disasters, the complexity of CBRN defense operations probably will increase in the future.

The development and use of CBRN technical equipment needs to be investigated from climate change's point of view as newly developed instruments should meet the changing environmental challenges. Wide range of research and development is necessary to fit up the soldiers with appropriate equipment, e.g. with more comfortable IPE's, with more durable weaponry and measurement units concerning operational temperature range and with more effective decontamination equipment and materials. At the same time, reconstruction of the CBRN training's system is also necessary [3].

# 5 Possibilities for Use of Renewable Energy Sources

There are many possibilities to modify the power sources currently running with fossil energy sources, to use alternative energy, in order to avoid limited operational time and to reach continuous service. It is important, as the renewable energy sources can be continuously at our service, and their use does not pollute the environment, that's why environmental protection regulations order the users to use them in increasing scale.

These energy sources gain their power from three different sources. The largest group of equipment uses the thermal energy coming from the Sun (solar cells, photovoltaic panels, wind, hydro) and partly from geothermal as secondary sources.

In the second group there are other types of energies, e.g. power stations using gravitational phenomena as the rise and fall of the see (tide) or heat of the Earth (geothermal energy).

We can put into the third group all of the energy sources available from the bodies of living organisms as biomass or any organic waste, which is indirectly solar energy as well.

Concerning the needs and specialties of the military there are a lot of limitations in the possible use of alternative energy sources both in Hungary and abroad. Such a limitation is the need of safety, which sometimes can overwrite even the best economic solutions.

By investigating all of the previously mentioned possibilities we can find during the application tests that the best opportunity is a device using directly the energy of the Sun and creating electric power: this is the photovoltaic solar panel. Its introduction means only a one time investment, transportation and installation cost, after putting into service it does not need any "fuel" and maintenance costs are very low.

In case of bases abroad, practically, solar panels are the only recommended device, because all the others have smaller savings and face larger safety problems. There can be many circumstances negatively affecting the performance of a renewable energy resource. Such problems for a solar cell occur when the sky is overcast for some time (rain, cloudy, shadows in general). In this case the solar panels cannot collect enough solar radiation so they cannot produce enough energy. This means that old, conventional equipment has to be kept, because they can be needed in the case of problems, but the main power supply must run with the alternative resources.

In case of domestic bases the types of applicable energy sources are much wider. Besides photovoltaic panels it is practical to use simple solar collection devices that generate hot water for heating and domestic purposes, wind power generators that can produce cheap electric energy, and concerning our countries exceptional geothermic capabilities, the use of geothermic energy can produce a enormous savings in money and in fossil energy supplies.

We should forget about the energy coming from biomass and organic waste processing. These products are not widely used in our country yet, so there are large capabilities in reserve. We need detailed plans that for currently existing domestic hot water and heating systems besides the use of solar panels and geothermic energy what are the costs if we produce energy from biomass (e.g. wood pellets) or the communal waste of the military base instead of making heat with the use of fossil fuel resources.

But overall, the installation of photovoltaic solar panels is the best solution for decreasing dependence on fossil energy consumption for military units concerning safety and economic issues [4].

# 6 Investigation of Military Clothing

During our research, we investigated all the types of military clothing (uniforms) of the Hungarian Defense Forces. We used a so called "thermal manikin" for modeling the thermal sensing behavior of a soldier. Winter and intermediate measures were made in a refrigerator container placed in front of the building of Sanitary Engineering Department of Pollack Mihály Technical and Informatics Faculty, University of Pécs (PTE). Summer measures were made in the Department's heat sense laboratory. During the experiments we measured the temperature and humidity of the environment, wind speed, and also made videos with a thermal camera. We recorded almost 20 million data, so we developed an EXCEL based program for data management and rapid calculations.

We made evaluation sheets containing data on the given clothing type, circumstances of the experiment, the so called CLO value of the clothing (thermal resistance, calculated from the results) and the specific and summarized performance of the thermal manikin. There are diagrams on the evaluation sheets for the specific and summarized performances of all designated body parts, for the further analysis of the single clothing parts.

During the experiments we calculated the specific and summarized performance of the thermal manikin for each clothing. As the soldier did the same task in every experiment (the metabolic equivalent, MET was the same), his specific performance should have been the same, independently of his clothing.

Our main findings were:

- In wintertime, in the case of fatigue and desert clothes, the specific performance of the thermal manikin was higher than the average specific performances of the measured clothes. The thermal manikin to be in thermal balance (PMV = 0, so he felt in comfort) needed extra performance. In this case PMV < 0, so the soldier felt cold. If there was some kind of activity in the clothing, and the volume of activity was 1.8–2.0 MET, the thermal balance was restored.
- In summertime, for case of fatigue and desert clothes the specific performance of the thermal manikin was higher than the average specific performances of the measured clothes. The thermal manikin to be in thermal balance (PMV = 0, so he felt in comfort) needed less performance. In this case PMV > 0, so the thermal manikin was hot. The thermal manikin cannot sweat, but the human body in this case can pass the surplus heat via sweating.
- In an intermediate thermal time (spring or autumn), for the case of fatigue and desert clothes the specific performance was around the average specific performances of the measured clothes, so the thermal balance was built in.
- For case of everyday and corporate clothing clothes the measured values were around the average specific performances, so the thermal balance was built in.
- For the case of wind the performance increases, in wintertime, either the CLO value of the clothing or the activity (MET value) has to be increased to achieve the thermal balance. In summertime, for the case of wind, the thermal manikin was in thermal balance.

The above findings refer to the whole clothing sets and do not contain the independent measures of each cloth parts.

Measured values with the thermal manikin in some cases were compared with same experiments with living humans. In these cases the results with the living humans verified our data with the thermal manikin. We have additional data concerning the separate cloth parts and these can be good starting values for further research.

The types of clothing were separated for several versions based on different ambient temperatures. The measures were made at a wide range of temperatures using the winter, intermediate and summer versions of the dresses. The winter period was defined from −20 to 0 °C, the intermediate from 0 to +15 °C, and the summer from +15 to +30 °C. So the lowest temperature during the experiments was −20 °C and the highest was +30 °C [5].

Military clothing should meet various requirements, the needs are very complex. In the following, we review the description made by Marianna Halász:

- Military clothing should give physiological comfort even during extreme climatic conditions and activities with very different intensity.
- Military clothing is a special smock. Its design should meet the planned activities and the normal labor safety regulations.
- As military activities are complex, the clothing should be comfortable, tight-fitting, light and flexible, should not hinder movements and body functions, and must be easy to take on and off.
- Must not be harmful for health, should not be chargeable with static electricity, should be mild for the skin, should not harm or scratch the skin during movements.
- Should meet the aesthetic requirements.
- Should be durable with proper mechanical strength, abrasion resistance, washing resistance and colorfastness, its material should resist heat, moisture, UV radiation and microorganisms (e.g. fungi).
- Military clothing should have a protective function. It should give the soldier possible protection against environmental harms, e.g. mechanical effects, sandstorms, parasites, insects, worms, snakes, chemicals (acids, bases, etc.), fire, UV radiation, electricity [6].

Physiological conformity of the military clothing is very important. One cannot wait a good performance from a soldier whose hands or legs are frozen in cold weather or cannot concentrate for his task because of a heat stress in warm weather. Physiological comfort should be guaranteed even in extreme weather conditions. At the same time, ergonomic, durability and protection requirements can be contrary to physiological aspects, so designing of military clothing needs trade-offs. The aim is improve physiological conformity of the military clothing while keeping the meet of the other requirements.

Military clothing should protect the soldier from any environmental harm as much as possible, at least as it is expected from non-special protective clothing. Unfortunately, these protective functions are working only when the body of the soldier is covered with the clothing. And in warm weather this protection and loose clothing are contrary wishes. The looser the clothing in warm weather, the weaker the protection it can provide.

The clothing must protect the soldier against mechanical effects. That's why so important the sufficient mechanical strength of the fabric, thread and other connection elements used for fabrication.

Soldiers often should fulfill their mission in natural environment where they can meet dangerous creatures. It is simpler to fight with visible ones than with small or hidden animals like parasites, insects, worms, snakes. Clothing should also provide protection against these ones. The protection against their penetration is given on one hand by the solid, strong and thick fabric of the clothes, and on the other hand by its proper design. From this point of view the clothing should be as closed as possible so the small animals cannot go under. So the clothes must be tight at its wrist, ankle and neck. And this closing is the most important during warm weather, because these animals are the most active at that time.

Increasing UV radiation is a serious threat coming with climate change. Fortunately, the skin is protected from the UV radiation even with normal clothing, but it is necessary that the whole body surface should be covered with clothes and a hat should be on the head of the soldier to protect him from direct sunlight.

Synthetic fabrics can be problems. Electrostatic charge can be dangerous if there are flammable materials in the vicinity, which can catch fire or even explode from a sparkle generating from static electricity. That's why important that the military clothing should not be chargeable with static electricity.

Military clothing should be durable in a reasonable extent. Durability can be characterized with mechanical strength, abrasion resistance, washing resistance and colorfastness, resistance against heat, moisture, UV radiation and microorganisms (e.g. fungi). It is important for the soldier's safety that his clothing can maintain its original parameters for a long time. Durability is affected even by the fabrication technology. It is important to use a strong, solid, colorfast thread with proper quality for sewing. Technology of sewing can affect durability and comfort for another way. For example, if a trousers is hardened with an extra layer of fabric at the most abrasive points e.g. at the knees and at the bottom, the result can be odd. The hardened part resists better against stretching and lasts longer, and double layer is better against abrasion, so it looks an advantage. But because of the double layer sewing gets thicker and becomes stiff and lifts up from the fabric, so it wears quicker, and it is definitely a disadvantage.

Physiological task of the clothing is to create a micro environment for the body suitable to maintain its permanent temperature and humidity of its vicinity. So the clothing should provide proper thermal insulation and a way for emission of inner humidity and at the same time protect from outer humidity.

Military clothing should protect the body from rain and other types of precipitation. The problem with precipitation is that if we do not defend against it, it can be absorbed by the clothes. So the clothing becomes heavier and evaporation removes heat from the body, resulting chilling and feeling cold. It can be a problem even in mild weather, because the humidity coming with precipitation can result in sub-cooling.

In extreme cold—below −5 °C—the main task of the clothing to keep the temperature of the body, to keep the body warm, to keep outside the humidity coming with the precipitation, and driving off the possibly creating sweat from the body surface. So the task is to solve the problem of simple, limited adjustment of the warming, precipitation protection and sweat driving, moisture permeability capabilities of the clothes during movement.

Thermal insulation capability is adjustable with the number of worn clothes layers. A bit more complex solution is the adjustment of the air permeability of the outer layer instead of the thickness of the thermal insulation layer. If we are hot and we increase the air permeability of the outer layer, we provide more intensive change for the closed air inside the thermal insulation and also more intensive leaving of humidity, but if we are cold, than decreasing the air permeability of the outer layer can prevent the change of the warmed air from inside.

This kind of regulation can be carried out by the use of one or several thin, light layers, that are restrictive for air or even impregnated against water and easy to take on and off. This solution lets the inner heat insulation layers remain unchanged during movement.

Experiments with living humans and the thermal manikin show together that arms and the back are often felt cold. In this case, the solution can be to put on another long sleeve upper layer under the outer layer. If a coat is part of the clothing, a warm interline should be used or it should be completed on the back and at the lower arms. The front of the coats are thicker, because they contain not only a silk interline, but also a stiffener calico interline. So the front of the coat is warm enough.

The head, hands and legs need special attention in case of heat insulation. The body primarily tries to keep the inner organs warm. If the heat capability of the body gets lower, than it decreases the heating of the less important peripherals as the hands and legs or ears and nose. So on one hand these body parts gets cold sooner, and on the other hand in case of feeling cold, the body decreases the heating of these parts. That's why it is so important to keep warm of these body parts. The soldier cannot wear simple boots in temperatures below 0 °C. It is important to have special winter boots. The issue with the hands is similar. One cannot make precise movements with frozen hands. There are some smart solutions for the hands to keep them warm. For example, normal five-finger gloves can be covered by extra one-finger gloves only on the fingers. In case the soldier needs his fingers, he just rolls up the one-finger glove part, if not, he rolls back the second layer. One thing is for sure: it is impossible to keep the hands warm in cold weather with the use of simple five-finger gloves. It is necessary to keep the fingers together to warm each other and it is only possible with the use of one-finger gloves.

Even the soldier dressed correctly warm; in case of staying for a long period on cold terrain without certain activities he will need some extra cold protection. In the absence of a tent or a sleeping-bag he should have a blanket similar to the one used by medics to keep the patient warm.

Sweat driving and moisture permeability capabilities are important even in case of winter clothing. Primarily, the regulation of the air permeability of the outer layer is important. This way the air exchange, moisture driving and heat insulation capabilities of the clothing will also become adjustable preventing the soldier from sweating. In case of sweating, it is subservient to drive the moisture off the body surface, because the cooling function will not work as the evaporation is blocked. If the sweat remains on the body surface, it can cause uncomfortable feeling. There are two chances. The first is to put on appropriate underwear to absorb moisture. But the moisture remains close to the body surface this way, causing subcooling in case of decreasing temperatures. The second possible solution is to create underwear from two layers. The inner layer (close to the body surface) should be made of a fibrous material such as polypropylene that cannot absorb water at all. This layer drives off the moisture from the body surface and transports it via capillaries to the other layer which can absorb it. This way the moisture cannot divert heat from the body for its evaporation. We can find this solution in many areas so probably it will work with military clothing also. Nevertheless, we would like to emphasize the danger, that in case of fire the polypropylene layer can melt causing serious injuries.

In case of extreme hot—above 30 °C—the situation is more difficult than in extreme cold. Unfortunately the soldier cannot take off his clothes while in duty. And he should wear his clothing to have its other functions besides physiological function especially the appropriate protection.

From physiological point of view, the main task of the clothing in warm weather is to make the free body heat transmission possible. Sweat is for cooling down the body, so it is a problem to drive off the moisture from the body surface.

In warm weather it is practical to wear clothes made of a textile with good air permeability and good moisture absorbance. The clothing should be made as light and loose as possible, for example from fabric made with panama-texturing. Loose clothing helps ventilation and evaporation of moisture this way helps cooling the body. This can be improved further with the use of additional net fabrics in the closed areas of the clothing for example on the side of the upper part and on the inside of the sleeves, but this could also lead to the decrease of protection. The ventilation of the hat can be improved with the implementation of small holes in the fabrics.

Use of cellular material in the clothing with its loose structure makes the evaporation of sweat and pass of moisture possible. It can be an advantage especially in case of using bulletproof vest with ceramic plate insertion in warm, summer weather. It is not necessary to mention the physiological discomfort features of the protecting vest. If the soldier wears appropriate underwear made of cellular material, it would improve the ventilation of the clothing to help the soldier feel more comfortable.

Soldier cannot stand without combat boots even in summertime. But the sweating of legs is inevitable in closed boots in summer. The discomfort can be decreased by the use of socks containing silver. Disinfective behavior of silver was well known to the ancients. Silver can be well applied to the textile so it cannot be removed by repeated washings. If silver applied in socks it can prevent the settle of fungi and the bad smell.

It is very important in case of clothing designed for warm weather to arrive at a compromise. To decrease a bit from the protection capabilities can improve the physiological behavior of the clothes. It is a very important decision how to keep the balance between.

There are many ways to increase the comfort of clothing. In some cases, it is only a question of attention and decision, but sometimes it is a financial issue. We did not mention earlier, such up-to-date solutions, as built-in heating and cooling modules operating with solar panels in clothes, gloves and boots, or special heat and flame resistant materials or aerogel, which is used by astronauts and polar researchers as heat insulation layer, as their implementation is not possible because of their high costs.

We have suggested solutions that can be implemented in present circumstances with limited financial capabilities and can definitely improve the comfort of military clothing. It is important to mention that we are all different in our levels of feeling comfortable. So our opinion is that we have to allow the soldiers to wear certain extra layers of clothes for their individual needs that are not visible from outside to have better level of comfort.

**Acknowledgements** We acknowledge the financial support of this work by the Hungarian State and the European Union under the TÁMOP-4.2.1B-11/2/KMR-2011-0001 project.

# References

1. Kovács, F.: The green barracks program. Hadtudomány, pp. 67–82. May 2013. http://www. mhtt.eu/hadtudomany/2013/eghajlatvaltozas.pdf, downloaded 20th April 2014
2. Hullám, I: Műveleti területek szélsőséges klimatikus tényezőinek hatása a katona pszichikai-mentális teljesítményére, pp. 83–100. Hadtudomány, May 2013. http://www.mhtt. eu/hadtudomany/2013/eghajlatvaltozas.pdf, downloaded 20th April 2014
3. Földi, L.: A klímaváltozás jelentette kihívások az ABV védelemben, pp. 101–116. Hadtudomány, May 2013. http://www.mhtt.eu/hadtudomany/2013/eghajlatvaltozas.pdf, downloaded 20th April 2014
4. Barbarics, T: Megújuló energiaforrások alkalmazása a hadseregben. Essay, manuscript at the author, pp. 67–116 (2014)
5. Magyar, Z., Révai, T., Lenkovics, L., Budulski, L.: Az éghajlatváltozás, a szélsőséges időjárás, különböző hőmérsékletek hatása a honvédségnél rendszeresített ruházatra. Research report, manuscript at the authors (2014)
6. Halász, M.: "Az éghajlatváltozás, a szélsőséges időjárás, különböző hőmérsékletek hatása a honvédségnél rendszeresített ruházatra" measures and subjective experiments of military clothing, data evaluation and suggestions for modification of military clothings. Research report, manuscript at the author (2014)

# Critical Transport Infrastructure Protection

## Results of a Supply Chain Security Research

**Attila Horváth and Zágon Csaba**

**Abstract** Risks associated with supply chains, belong to the transport sector of critical infrastructure protection and require appropriate responses and preliminary measures respectively. The analysis of the focus and objective of the particular research, points out the need for complex approaches and the expectations for the multipurpose applicability of the results. Starting from the threats, this successful research has identified the needs for an advanced network analysis to reach door-to-door security along the entire length of the supply chain.

**Keywords** Critical infrastructure protection · Door-to-door security · Network analysis · Supply chain security

## 1 Introduction

The Critical Transport Infrastructure Protection Priority Research Area was carried out as part of the Project #TÁMOP-4.2.b-11/2/KMR-001. The research team enjoyed full autonomy since the inception of the sub-program in line with the best practices of the leading countries of the topic researchers. The main objectives the research program were determined as follows:

1. Conceptualization of critical infrastructure protection and the possible adaptations of foreign experiences respectively.

A. Horváth (✉)
Department for Operational Logistics, Military Logistics Institute,
National University of Public Service, Budapest, Hungary
e-mail: horvath.attila@uni-nke.hu

Z. Csaba
Doctoral School for Military Sciences, National University of Public Service,
Budapest, Hungary
e-mail: csabaz@dravanet.hu

© Springer International Publishing Switzerland 2016
L. Nádai and J. Padányi (eds.), *Critical Infrastructure Protection Research*,
Topics in Intelligent Engineering and Informatics 12,
DOI 10.1007/978-3-319-28091-2_8

2. Analysis of the spheres, sectors, institutions and legal frameworks of the critical infrastructure protection according to international examples.
3. The role of traffic systems in the critical infrastructure protection.
4. Allocation of the assignments and methodologies for each traffic mode (air, land, rail, sea, inland waterways, pipeline and cable transports) in critical infrastructure protection.
5. Vulnerability analysis of the infrastructure elements of traffic modes and sectors (e.g. carriageways and tracks, structures, terminals, vehicles, facilities and equipment, control systems etc.) in the European Union, as well as, in Hungary. Possible risk areas for the analysis may be terrorism, natural and human-made disasters, technological hazards, sabotage, anomalies, incidents, accidents and global warming related environmental effects.
6. Analysis of the interactions between traffic systems from a critical infrastructure protection point of view.
7. The interdependency of the traffic systems on the critical infrastructure protection of the energetic networks, info-communication technologies, food and water supplies, financial systems, public administrations and government sectors.
8. The interdependency of the security and defense sector on the critical traffic infrastructure protection.
9. Security and safety researches at the priority areas of the supply chains and logistics networks.

The research team was composed of nine academic researchers, four Ph.D. candidates and a M.Sc. student. Introducing the facts and figures concerning the key research indicators, we may mention two strategic and a practical research report, 17 articles published in academic journals as well as 28 conference presentations. Two books containing 15 and 8 chapters were considered as one of the main achievements of the project. Both books were issued in print and e-book versions.

At the inception, the research priorities were supplemented with supply chain security, and the public awareness in emergency situations and their preparedness capabilities. These amendments were proven important from public expectations as well as the research objectives points of view [1]. The dependency on technology is increasing continuously in the postmodern age and the issue of security becomes more and more a priority either from its economic, socio- or other approaches. This study will sum up the research results of the supply chain security area.

## 2 The Result of a Globalised World: The Globalised Supply Chains

In the age of a global world, no economy can function without the security of supply chains. A permanent barrier in the supply chain may paralyze production, distribution and sales processes a few thousand kilometers away. At the beginning of the research, we set up the hypothesis that there is no sufficient what to discuss about the protection of the traffic systems within the greater phenomena called critical infrastructure protection.

In the recent decades, the supply chains became a part of life for societies and even for the individual members. The energy sources, clothing, technical equipment, a good portion of food products—consumed by everyday people—were elaborated in countries far away. These products were manufactured, distributed and sold through complicated systems until they arrive at the consumers.

The previously unimaginable integration could not have taken place without the spread of a global economic approach, new manufacturing techniques, the developments in transportation modes and methods of logistics, the increased use of IT systems, etc. New economic trends appeared, certain technologic advances became integral parts of the system while consumer habits were also changed. These achievements led to the spread of the advanced supply chain management [2].

## 2.1 Defining a Concept

Just as for the terms of security, terrorism and logistics, we can find hundreds of conceptual definitions for the supply chain that meet scientific standards. The common feature of supply chain related concepts is that the circulation of materials and products is always associated with the flow of information—this approach appears everywhere [3]. Three main sections may be separated in the supply chain for the material and information flow related processes such as the procurement, the manufacturing and the customer services [4]. When a product that is developed in Europe or in the United States, manufactured in China from elements originating from almost all over the world, it becomes a part of the global sales process assuming well coordinated processes. In such a complex and networked system, the co-operations have to cross over the traditional corporate structures [5]. The supply chains are one of the occurrences of globalization. They have been created and at the same time maintained, by the internal values of this phenomenon. To be protected against the harmful effects of the global supply chain may be efficient if its operational mechanism is well considered and their opportunities are exploited [6].

## 2.2 Challenges of Our Times

The protection of sources of supply, manufacturing plants, warehouses, commercial facilities cannot be considered a new issue. Throughout history, after the formation of organized human societies the protection of these facilities has always been priority to the stakeholders and to the states, as well. Both the security of economy and trade became a strategic issue from public administrative and military concerns. In the Cold War period, the opposing parties were evaluated as secure in military terms almost exclusively. After the disintegration of the bipolar world, security studies have become more open to other aspects of the real world other than the military and foreign policy problems, which were in fact, their primary focus at that time [7]. However, it would be a mistake to state that the altering "new security challenges, risks and threats" (as these phenomenon are often specified by security experts), such as terrorism, the mass migration, drug trafficking and other forms of organized cross-border crime, etc. have not previously been posed threat to the societies and the states [8]. The post-modern society and the evolution of civilization entail numerous challenges that pose serious risks to the global, continental, sub-regional security. With the increasing environmental burden shifting, the negative effects of overpopulation and global warming have already impacted on food production and supply chain events. This made the elaboration of strategies and action plans necessary to facilitate the handling of such challenges.

## 2.3 Examinations in the Focus

Our recent time pre-eminent thinkers are preoccupied with the examination of post-modern societal and technological developments related phenomena. The French philosopher Jean Baudrillard published several studies in preparation to the millennium dealing with the phenomena emerged in the postmodern era. In his works the resources of security risks were also affected. He saw the primary problem sources of the inequality in controversial political and economic relationship between the West and the Middle East, Southeast Asia, Africa and Latin America. However he broke with the geopolitical context that was traditionally interpreted as a reference model "North versus South" [9]. The author interprets the term of globalism in the opposition and the mutual interdependency between the West and the less developed parts of the world. Baudrillard's novel approach is thought provoking in many aspects.

## 3 Geopolitics and General Security Contexts

Since 1990, we have witnessed a change that amended the economic and security dimensions and probably reshapes the geopolitical map of the world. There is no doubt that the "BRICS countries" have a decisive role in world economic developments. The Goldman Sachs introduced term incorporates countries might join the group of economically developed countries in a longer perspective and includes Brazil, Russia, India, China and South Africa [10]. These countries have already proven their unavoidable role in transportation of certain groups of goods throughout the supply chains.

## 3.1 Mandatory Network Analysis

At this point we need to turn to the networks. BRICS countries may be found among the most frequent container ports of the word and they form a vital network of supply chains with the most developed countries added with certain choke points, such as Singapore. The graph theory and network analysis calls last the mentioned type of nodes connecting centrality, because they provide not simply links between two important groups of nodes, but these types of connections quantify the number of times a node acts as a bridge along the shortest path.

If we want to ensure door-to-door security in the networks of supply chains, we need to turn to recent achievements of network analysis. It is absolutely necessary to define those parts of the network, where security measures would be the most affected. In other words, we have to point out the most vulnerable points on the shield of defense and target our limited capabilities accordingly to protect the network as effectively as possible. From this aspect, connecting nodes have just as high relevance, if not even more, than the in-degree and out-degree nodes that are used as the network's most frequent input and output points.

If supply chains form a scale-free network [11], the network has a high degree of fault tolerance and may not be disrupted by random attacks with much ease, due to the proven character of such network types. In order to disrupt it, targeted attacks have to be triggered at those nodes, which have the highest number of links. This knowledge proves useful in either destroying and protecting the network.

## 3.2 Global Network Patterns

The global economic character may not simply be considered, because of capital investments and settlements of economic operators favoring those areas, where the necessary conditions are provided for a prosperous operation. The sources of raw materials, the repository areas, production sites, sales and distribution systems are often located thousands of kilometers away from each other. How to guarantee security along with the transportation routes between these points? Supply chain security is therefore, a serious challenge for global, continental, regional, and national level [8]. The "Think globally, act locally" slogan stands for the security efforts in the supply chains as well, and leads us to better solutions.

The Icelandic volcano Eyjafjallajökull eruption case in April 2010 supports the above statements. Such natural disasters are frequently occurring and usually result in no fatalities in Iceland. But the secondary impact were even worse than the direct primer local consequences, because a serious disruption occurred in the air transport system all over Western Europe. The scope of the damage should not have been estimated from the missing airfares solely, but the secondary consequences e.g. the loss of business travels and the delays of air consignments also had to be considered.

The Japanese, Fukushima nuclear power plant disaster, was caused by an earthquake and a subsequent tsunami on 11 March 2011. It is less known, that leading Japanese enterprises, in the global IT sector were supplied with an inadequate amount of electricity as a consequences of the disaster and therefore they could not maintain their continuous production. If the blackout had not been solved in a relatively narrow time frame, this would have caused high risks for the global IT industry, because they would sustain a serious disruption. If this occurs, it would further deepen the worldwide economic crisis, because of the IT sectors' high significance in the global economy [8].

Even strategically relevant terrorist attacks can rarely incur extensive direct consequences, but their indirect effects can easily affect the security of the supply chains. Due to the closure of seaports and airports, serious disruptions occurred in American foreign trade after the series of terrorist attacks on 11 September 2001 [12]. Thus, the terror attacks deemed the most serious, pointed out that terrorism and the countermeasures taken against their consequences may distract the supply chain and, in severe cases, the whole economy. This represents risk factors other than terrorism that may affect a network of properly operating supply chains. As an effect of the interdependency, the serious changes (such as the above mentioned forms of shocks) would necessarily impact on another infrastructures.

## 3.3  Wave-Like Diffusions

It is still an unsupported assumption that a serious impact, such as a shutdown of infrastructures from an unexpected event, will produce a subsequent burden of transfer to other infrastructures, in a wave-like manner. The first event will trigger the strongest effect, that is followed by other events, repeatedly. The repeated occurrences will exhibit declining effects. This would raise more questions in the prevention, minimization and recovery of damages if this theory is how cases develop. Moreover, this also suggests, deciding on how many repeated occurrences should be taken into consideration for a more precise estimation of damages, as well as at the use of countermeasures and organized recovery [13].

## 4  What the Security Studies Should Cover?

As we revealed previously, pre-eminent international thinkers had serious debates on what questions should be involved in the security studies and the research respectively. In this regard, Copenhagen School researchers introduced an internationally acceptable recommendation. Accordingly, those threats may include changed approaches to security, which concerns existence-threatening risks and the respective risk management may take extraordinary measures as necessary [14].

The future is not simply ranking among the different risk factors, but the real threat assessments. We need to point out that the supply chain forms a multi-player and many-factor system, therefore, its vulnerability also depends on many other variables. The only security of the supply chain can be understood as an overall idea, which encompasses the raw material yields, to consumption or recycling. This is expressed by the neat term, "door-to-door security". However, due to large spatial differences in the supply chain the risks may differ enormously. The new American approach to risk management, broke the primacy of terrorism, among the threats that were interpreted, in general, for the transport and logistics sectors, affecting mainly public transport systems. Due to the appreciation of the international economic role of the supply chains, we need to consider the risks of the freight transport systems [15]. Although the attacks against logistic sites and transport vehicles were not so frequent in the history of terrorism, these threats must seriously be considered.

## 5 Examples to Follow

There is a repeated dispute between concerned parties as to how far the freedom of entrepreneurship would go and where from the strict rules of security should prevail. Issues of supply chain security, the complex nature of potential risks and impacts of the possible occurrences of extreme events require close cooperation of the participating organizations. This problem goes far beyond the scope of critical infrastructure protection in the transport sector. For the sake of a better cooperation, the introduction of clear international regulations are proposed—preferably with market and financial consequences—that would take safety aspects into account, instead of driving down the expenses of production at the expense of safety. The industrial and the agricultural sectors are of those where cooperation should be ensured for the higher security of the supply chain. The food-related scandals of the recent years pointed out that security must be interpreted for the entire food supply chain involving production facilities, processing plants, commercial units and end users [16]. The mutually agreed procedures highly facilitate the development and control of the rules concerning logistics and transportation services between the two sectors. Collaboration with the participants of the food supply chain would benefit and allow examples to follow for any operational sectors of the supply chains. The lessons learnt accordingly, would also foster the proper communication of incidents [16]. The experiences of recent years in this area abound in samples that can be and should be taken over, or also those ones that can already be ruled out.

Keeping regulations provides some guarantee, but the real step forward would be building up such security systems, which cover all activity forms of the firm comprehensively, instead of fence construction, installation of CCTV systems and the use of security staff, that companies usually do. Providing legal guarantees and the risk of market loses, together, would not even be sufficient motivations; security should become the part of the organizational culture.

It is not appropriate to assess risk factors separately in such complex systems like the supply chains. We follow those experts who claim that these systems should be evaluated with external and internal interactions of risk factors in a complex manner [17]. Both the ignorance of the interdependencies [18] and the overemphasis of certain security factors would result in errors within the assessment [19]. The enlarged threat of terrorist attacks has increased the complexity of prevention of the incidents [20].

# 6 Summary

The security of supply chains and the critical infrastructure protection are overlapping territories. The prevention and if the incidents have already occurred, the restoration plays a vital role in both areas and preparedness is always a key. We have seen that unpreparedness was often covered up by targeted PR, filling gaps like that, never succeeded [21]. The guarantee for the success is cooperation of the economic operators, facility owners, governments, their respective agencies and international organizations. This cooperation can even occur across continents, if the character of the supply chain requires [22]. We consider high a responsibility of researchers in supply chain security, because the research is still on-going in many countries and the regulations and means of co-operations are under constant development.

**Acknowledgements** We acknowledge the financial support of this work by the Hungarian State and the European Union under the TÁMOP-4.2.1B-11/2/KMR-2011-0001 project.

# References

1. Macaulay, T.: Critical Infrastructure: Understanding Its Component Parts, Vulnerabilities, Operating Risks, and Interdependencies. CRC Press, London (2008)
2. Knoll, I.: Logisztika-gazdaság-társadalom. Kovásznai Kiadó, Budapest (2002)
3. Szegedi, Z.: Ellátási lánc-menedzsment. Kossuth Kiadó, Budapest (2012)
4. Bowersox, D.J., Closs, D.J., Cooper, M.B., Bowersox, J.C.: Supply Chain Logistics Management, 4th edn. Mcgraw-Hill International Edition, New York (2013)
5. Szegedi, Z., Prezinszki, J.: Logisztikai menedzsment. Harmadik kiadás. Kossuth Kiadó, Budapest (2005)
6. Horváth, A., Csaba, Z.: On the vulnerability and reliability of towns and cities. In: Csapó, T., Balogh, A. (eds.) Development of the Settlement Network in the Central European Countries, pp. 299–312. Springer, Heidelberg (2013)
7. Walt, S.M.: A biztonsági tanulmányok reneszánsza. In: László, P. (szerk.) Nemzetközi biztonsági tanulmányok, pp. 9–52. Zrínyi Kiadó, Budapest (2006)
8. Horváth, A.: A kritikus infrastruktúra védelem komplex értelmezésének szükségessége. In: Horváth, A. (szerk.) Fejezetek a kritikus infrastruktúra védelemből. I. kötet, pp. 25–48. Magyar Hadtudományi Társaság, Budapest (2013)
9. Horrock, C.: Baudlliard és a millennium, Alexandra, Pécs (2008). ISBN 9633684595

10. Lakatos, J.: BRICS avagy nagyhatalmak tranzitban. Méltányosság Politikaelemző Központ Budapest (2011). http://www.meltanyossag.hu/files/meltany/imce/doc/ip-brric-110412.pdf. Downloaded 18 Mar 2012
11. Barabási, A-L.: Behálózva – A hálózatok új tudománya. Budapest, Magyar Könyvklub (2003)
12. Cook, T.A.: Managing Global Supply Chains. Auerbach Publications, Taylor & Francis Group, London (2008)
13. Csaba, Z.: A tengeri konténeres áruszállítás biztonsága. In: Horváth, A. (szerk.) Fejezetek a kritikus infrastruktúra védelemből. I. kötet, pp. 133–166. Magyar Hadtudományi Társaság, Budapest, (2013)
14. Buzan, B., Wæver, O., de Wilde, J.: A biztonsági elemzés új keretei. In: László, P. (szerk.) Nemzetközi biztonsági tanulmányok, pp. 53–112. Zrínyi Kiadó, Budapest (2006)
15. Foltin, P.: Security of logistics chains against terrorist threats. In: 17th International Conference the Knowledge Based Organization. Nicolae Balcescu Land Forces Academy, Sibiu (Romania), 24–26 Nov 2011. Conference Proceeding Management and Military Science, pp. 100–105
16. Kasza, Gy., Surányi, J., Lakner, Z., Bódi, B., Deák, F., Horváth, A., Mészáros, L., Szántó, A., Danczák, I.: Rendkívüli helyzetek és kezelésük az élelmiszer-kereskedelemben - irányelvek tapasztalatok. Élelmiszer Vizsgálati Közlemények 68. évfolyam (2012), 3–4 szám, pp. 101–117
17. Asbjørnslett, B.E.: Assessing the vulnerability of supply chains. In: Zsidisin, G.A., Ritchie, B. (eds.) Supply Chain Risk, pp. 15–32. Springer, New York (2009)
18. Dani, S.: Predicting and managing supply chain risks. In: Zsidisin, G.A., Ritchie, B. (eds.) Supply Chain Risk, pp. 53–66. Springer, New York (2009)
19. Haimes, Y.,: In: Goetz, E., Shenoi, S. (eds.) Risk Analysis in Interdepedent Infrastructures, pp. 297–310. Springer, New York (2008)
20. Horváth, A.: Terrorizmus és térjellemzők a létfontosságú rendszerek védelmében. In: Horváth, A., Bányász, P., Orbók, Á. (szerk.) Fejezetek a létfontosságú közlekedesi rendszerelemek védelmének aktuális kérdéseiről. Tanulmánykötet, pp. 7–26. Nemzeti Közszolgálati Egyetem, Budapest (2014)
21. Kasza, Gy., Lakner, Z.: The bird flu in mind of Hungarian consumers-lessons and experiences of a direct-question survey. Acta Agrar. Kaposvariensis 10(2), 229–237 (2006)
22. Szyliowicz, J.S.: International transportation security. Rev. Policy Res. 21(3), 351–368 (2004)

# Evaluation of Differences in the Estimated Recrystallized Volume Using Different Methods Based on EBSD Data

András Mucsi and Péter Varga

**Abstract** There are several methods that have been developed in the last decade which use the data extracted from EBDS measurement for the purpose of determining the recrystallized volume of deformed and annealed metals. Two general ideas could be identified: calculations based on the quality of indexing of the scanned points and calculations relying on the differences in crystallographic orientations. First, a simple method is presented utilizing the Image Quality (IQ) indexing of the EBSD data. Estimations were also made analyzing the orientation maps resulted from the same measurements. In the final section the applicability of the IQ based method is discussed in comparison to the misorientation (MO) technique. Based on the accuracy of the measured and calculated results the proposed method seems useful although both procedure offer unreliable estimations at low recrystallization rates.

**Keywords** Recrystallization · EBSD

## 1 Introduction

There are a respectable number of factors hindering the precise quantitative analysis of the recrystallized volume of metallic products. To overcome this matter a number of techniques have been developed. Over the course of the 90s the rapid evolution of the microelectronics and the measurement processes related to the backscattering of electrons, namely the electron backscatter diffraction (EBSD) technique, enabled the establishment of new powerful evaluating methods.

A. Mucsi · P. Varga (✉)
Donát Bánki Faculty of Mechanical and Safety Engineering, Óbuda University,
Budapest, Hungary
e-mail: varga.peter@bgk.uni-obuda.hu

A. Mucsi
e-mail: mucsi.andras@bgk.uni-obuda.hu

© Springer International Publishing Switzerland 2016        101
L. Nádai and J. Padányi (eds.), *Critical Infrastructure Protection Research*,
Topics in Intelligent Engineering and Informatics 12,
DOI 10.1007/978-3-319-28091-2_9

From the increasing number of studies of the last decade two general methods can be classified based on the set of information used for determining the recrystallized fraction: Image Quality (IQ) and misorientation (MO) based. As for the first case recrystallized and non-recrystallized grains are characterized differentiating in the quality of Kikuchi patterns. The latter method characterizes the fractions of the scanned area by the orientation differences between the scanned points. Numerous techniques were developed in order to increase the reliability of the results which comes from different factors.

Early studies have found that the IQ indices correspond well to the recrystallized state of the material [1]. It is proven however that the IQ index is subjected to the quality of the surface, the microscope and imaging system used, the crystal structure and orientation and the chemical composition [2–4]. Thus several methods were developed to overcome these drawbacks [5, 6]. Tarasiuk et al. [7] applied a differential approach over a threshold value to distinguish the recrystallized volume. However, results from IQ based methods seem to be ambiguous [3]. Of course misorientation based techniques also have their drawbacks. The step size and the angular resolution have a major impact on the accuracy of the calculated recrystallized fraction [3, 8]. There are a number of methods to correctly determine the recrystallization parameters [5, 9], including the three parallel line scans technique [10] and the procedure based on the point-to-point misorientation within grains [2]. A recent study from Guilin Wu and Jensen [8] gives an automated tool to determine the recrystallized fraction which proven to be reliable if carefully presented [3, 8].

On the following pages a simple technique for determining the recrystallized volume from EBSD IQ values is introduced. The results are then compared to the ones gained from the aforementioned automated process based on the EBSD MO maps.

## 2  Experimental and Methods

For examination purposes pure copper (99.999 %) were cold rolled in five equal consecutive cycles to a true strain of 1.88 which is defined as

$$\lambda_{true} = \sqrt{\frac{2}{3} \cdot (\lambda_a^2 + \lambda_b^2 + \lambda_c^2)}, \tag{1}$$

where $\lambda_a$, $\lambda_b$ and $\lambda_c$ are logarithmic strains given as follows

$$\lambda_a = ln\left(\frac{a}{a_0}\right), \lambda_b = ln\left(\frac{b}{b_0}\right), \lambda_c = -(\lambda_a + \lambda_b). \tag{2}$$

The thickness and width of the plate before rolling are given as $a_0$ and $b_0$ respectively. Cylindrical test samples from the centerline of the rolled sheet with a

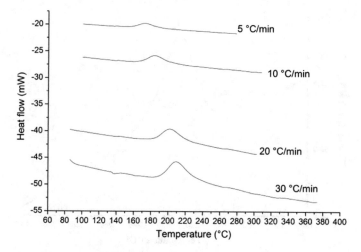

**Fig. 1** DSC thermograms recorded at different heating rates

diameter of 5.9 mm, a height of 1.5 mm and a mass of around 350 mg were prepared for annealing.

For annealing purposes Perkin-Elmer DSC-8000 differential scanning calorimeter (DSC) was used applying four different heating rates: 5, 10, 20 and 30 °C/min. The specimens were heated in alumina crucibles under the protection of N2 atmosphere with a gas flow rate of 20 ml/min. To determine the temperature intervals where the recrystallization will occur test measurements were made at a range of 20–350 °C (Fig. 1). Each thermogram contains an exothermic peak. The area under the peaks proportional to the released heat which is associated with the recrystallization process [11]. According to Fig. 2 the recrystallization starts at Ts and finishes at Tf temperatures. The related IQ and MO maps in Fig. 3 show the evolution of the microstructure at Ts, Tp (peak) and Tf temperatures.

The cold rolled specimens prepared for the purpose of evaluating the recrystallized fraction were heated to different temperatures between Ts and Tf then quenched in water thus producing partially recrystallized microstructure. To obtain the IQ and MO maps EBSD measurements were performed using a Philips XL 30 SEM. On each specimen three areas of 300 × 300 μm were scanned at a step size of 2 μm.

## 2.1 Determining the Recrystallized Fraction from IQ Maps

The analysis of the IQ maps provided by the OIM software consists of three steps:

- Saving the maps from the OIM software in bitmap format
- Loading these maps in an image analyzer software
- Subjectively designating the boundaries of the recrystallized areas by applying a threshold value on the grayscale image

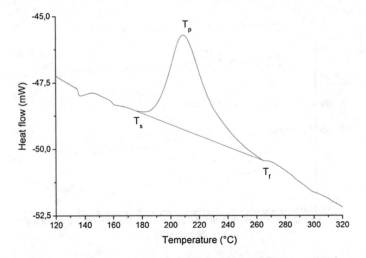

**Fig. 2** The DSC curve peak at a heating rate of 30 °C/min

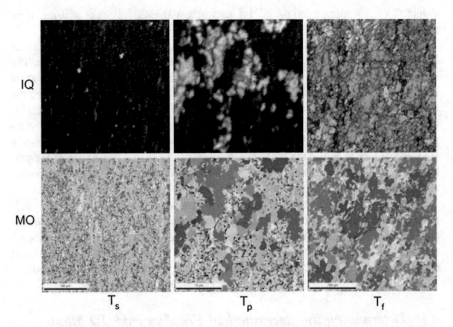

**Fig. 3** The evolution of the microstructure during recrystallization at a heating rate of 30 °C/min

The image of the IQ map consists of pixels with different shades of gray in accordance with the IQ indexing of the OIM software. The recrystallized fraction was calculated from

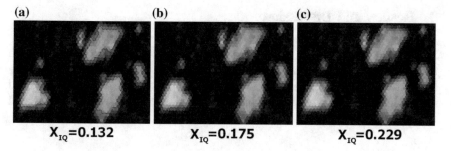

**Fig. 4** The dependence of the recrystallized fraction on the changes of threshold limits

$$X_{IQ} = \frac{N_R}{N_T},$$ (3)

where $N_R$ is the number of pixels considered to be recrystallized, $N_T$ is the total number of pixels the image contains. For each scanned area the selection of the set of pixels considered to be recrystallized was repeated three times thus resulting nine $X_{IQ}$ values for a specimen. In each selection procedure a grayscale value was randomly selected. The boundaries of the recrystallized areas were subjectively designated by manually changing the threshold limits of the deviation from the grayscale value. Figure 4 shows the changes in $X_{IQ}$ values related to different threshold limits.

## 2.2 Evaluating the Recrystallized Fraction Relying on the MO Data

The MO based evaluation was done on the basis of an algorithm developed by Guilin Wu and Jensen [8], and was implemented with their in-house software called DRG. MO maps of the same scanned areas as for the IQ based method were analyzed. According to the algorithm a feature to be considered as a recrystallized grain is assumed to obey several criteria: it should be a coherent area with a maximum given misorientation, its size should exceed a certain threshold level and it should be surrounded at least partly by high angle grain boundaries. Thus several parameters should be considered and set in the software. Out of these parameters only the value of the maximum grain misorientation angle (GMA) [2] was set to three different values, the rest were left as the software's default, resulting three calculation runs on each MO maps.

## 3    Results and Discussion

To evaluate the accuracy of the recrystallized fractions calculated with both
methods ($X_{IQ}$, $X_{MO}$) the standard empirical deviation of these values and the relative error of the measurements were determined. The two set of relative errors and
recrystallized fractions were then compared to validate the results of the IQ based
method.

### 3.1    Accuracy of the IQ Based Method

For each specimen, differentiated by the heating rate and the quenching temperature, the standard empirical deviation of the calculated recrystallized fractions
($\sigma(X_{IQ})$) was determined as follows:

$$\sigma(X_{IQ}) = \sqrt{\frac{\sum_{i=1}^{n}\left(X_{IQ,i} - \overline{X_{IQ}}\right)^2}{n-1}}, \tag{4}$$

where n is the number of measurements, $X_{IQ}$, i is the ith and $\left(\overline{X_{IQ}}\right)$ is the average
value of the determined $X_{IQ}$ values.

The highest deviations are related to the values of recrystallized fractions around
0.5, according to Fig. 5. The nature of the subjective selection of the recrystallized
area dictates that the more uniform the distribution of the grayscale, and thus the IQ,
values are over the whole scale the higher the deviations between the selected areas

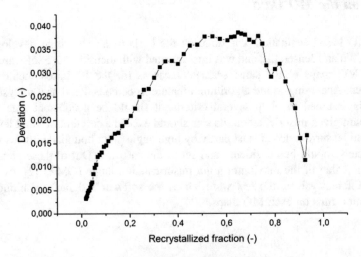

**Fig. 5** The deviation $\sigma(X_{IQ})$ as a function of the average value $\left(\overline{X_{IQ}}\right)$ of the calculated
recrystallized fractions

will be. These uniform distributions are typical of the partially recrystallized states [7]. The relative error of the estimation of recrystallized fractions from IQ maps is defined here as:

$$\Delta X_{IQ} = \frac{\sigma(X_{IQ})}{X_{IQ}}. \tag{5}$$

The changes in the relative error $\Delta X_{IQ}$ over the course of the recrystallization process are plotted in Fig. 7.

## 3.2 Accuracy of the MO Based Method

To define a set of data points in the MO map to be a recrystallized grain a maximum allowed misorientation angle has to be considered. Among other factors the maximum grain misorientation angle (GMA) has a high impact on the evaluated recrystallized fraction. The stricter the threshold limit for the maximum GMA the lower the value of the recrystallized fraction will be [2, 8]. According to Davies et al. [2] in a fully recrystallized grain the maximum GMA is between 0.7 and 0.8°.

To understand its effect some samples were analyzed with the DRG software covering a wide range ($\sim$0.4 to 2.5°) of GMA. At small average transformed fractions this method gives a large deviation in the value of the estimated recrystallized fraction over the whole examined GMA range (Fig. 6a, b). However at larger average transformed fractions ($\overline{X_{MO}} > 0.4$) the estimated values deviate much less at higher than 1° of GMA values (Fig. 6c, d).

Taking into consideration the results from previous studies and from our measurements 0.9, 1 and 1.2° were selected as parameters for the maximum GMA upon calculating the recrystallized fraction. This, multiplied by the three scanned area, resulted nine values of recrystallized fraction ($X_{MO}$) per sample which were calculated as [8]

$$X_{MO} = \frac{A_R}{A_T}, \tag{6}$$

where AR is the area of the MO map considered to be recrystallized by DRG and AT is the total area of the MO map. The values of the standard empirical deviation for each specimen ($\sigma(X_{MO})$) and The relative error ($\Delta X_{MO}$) regarding the estimations of recrystallized volumes were calculated in a similar manner as of the ones on the basis of IQ data. This means that

$$\sigma(X_{MO}) = \sqrt{\frac{\sum_{i=1}^{n} \left(X_{MO,i} - \overline{X_{MO}}\right)^2}{n-1}} \tag{7}$$

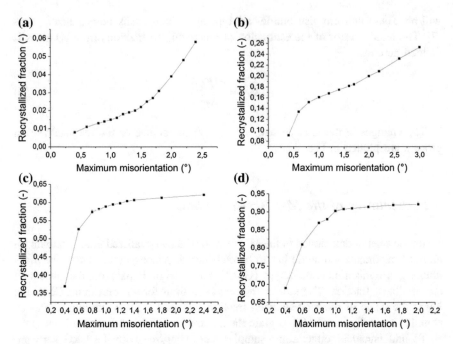

**Fig. 6** Recrystallized fraction as a function of maximum GMA ($\overline{X_{MO}} \approx$ **a** 0.02; **b** 0.17; **c** 0.6; **d** 0.92)

and

$$\Delta X_{MO} = \frac{\sigma(X_{MO})}{X_{MO}}. \tag{8}$$

where n is the number of measurements, $X_{MO}$, i is the ith and $(\overline{X_{MO}})$ is the average value of the determined $X_{MO}$ values.

## 3.3 Comparison of the IQ and MO Based Methods

Looking at the results of the technique based on the analysis of IQ index distribution proposed by Tarasiuk et al. the degree of recrystallization at low rates ($\sim 1$ %) can differ tenfold from the calculated value. The difference is reducing to $\sim 25$ % as the recrystallization rate increases to around 90 % [7]. Recent studies imply that the MO based evaluations can be reliable given that they are carefully implemented [2, 3].

Not surprisingly we found that the accuracy of both methods is highly dependent on the recrystallized volume. Figure 7 shows the concurrent values of relative errors $\Delta X_{IQ}$ and $\Delta X_{MO}$ as a function of the estimated recrystallized volume ($\overline{X_{MO}}$). As the

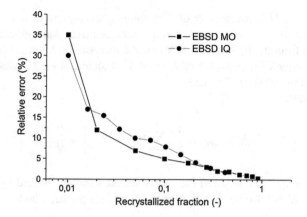

**Fig. 7** Changes in the relative errors $\Delta X_{IQ}$ and $\Delta X_{MO}$ over the course of the recrystallization process $(\overline{X_{MO}})$

recrystallized volume reaches the ratio of roughly 2 % the relative error of the MO based evaluations shrinks below 10 %. The error of the estimations based on the IQ indexing is around 3 to 5 % greater on the recrystallization range up to 20 %. Above 20 % the results from the two methods are correspond well. This correlation is shown from another aspect in Fig. 8 where the recrystallized fractions calculated with the two methods are plotted against each other. To define the correlation

$$C = \sum_{i=1}^{N} \left| X'_{IQ} - X'_{MO} \right|, \tag{9}$$

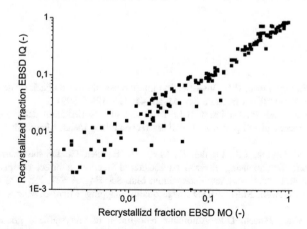

**Fig. 8** Correlation between the average recrystallized fraction determined on the basis of IQ map and MO data $(\overline{X_{IQ}}, \overline{X_{MO}})$

was used, where N is the number of data points (each represents one scanned area on the specimens), X'IQ is the average recrystallized fraction calculated per scanned area from the IQ map and X'MO is the recrystallized volume from the MO data for the value of maximum GMA set to 1°. Calculated for 156 data points we obtained a correlation of C = 4.53.

The formula:

$$\Delta X_{IQ-MO} = \frac{1}{N} \cdot \sum_{i=1}^{N} \left| X'_{IQ} - X'_{MO} \right| = \frac{C}{N} \tag{10}$$

gives the average error between the recrystallized fractions defined by the IQ and MO based methods for the whole population of data points, which is found to be 0.03.

# 4   Conclusions

Despite its simplicity, the results from the proposed method, based on the analysis of the IQ map, are in good agreement with the estimations constructed from the MO data. It is found that evaluations based on EBSD data give unreliable estimations below ~ 10 % of recrystallized volume. The calculations are still to be compared to results from other techniques which detect the degree of recrystallization based on different effects.

**Acknowledgements** This research has been supported by the assistance of the European Union, by the co-financing of the European Social Fund (TÁMOP-4.2.1.B-11/2/KMR-2011-0001). The authors are also grateful to the Department of Materials Science and Engineering of Budapest University of Technology and Economics to made the EBSD measurements possible.

# References

1. Black, M.P., Higginson, R.L.: Investigation into the use of electron back scattered diffraction to measure recrystallised fraction. Scr. Mater. **41**, 125–129 (1999)
2. Lu, H., Sivaprasad, P., Davies, C.H.J.: Treatment of misorientation data to determine the fraction of recrystallized grains in a partially recrystallized metal. Mater. Char. **51**, 293–300 (2003)
3. Dziaszyk, S., Payton, E.J., Friedel, F., Marx, V., Eggeler, G.: On the characterization of recrystallized fraction using electron backscatter diffraction: a direct comparison to local hardness in an if steel using nanoindentation. Mat. Sci. Eng. A **527**, 7854–7864 (2010)
4. Wright, S.I., Nowell, M.M.: EBSD image quality mapping. Microsc. Microanal. **12**, 72–84 (2006)
5. Lassen, N.C.K., Jensen, D.J.: Automatic recognition of recrystallized grains in partially recrystallized samples from crystal orientation maps. In: Szpunar, J. (ed.) Proceedings of ICOTOM 12, vol. 2, pp. 854–859. NRC Research Press, Ottawa (1999)

6. Humphreys, F.J.: Quantitative metallography by electron backscattered diffraction. J. Microsc. **195**, 170–185 (1999)
7. Tarasiuk, J., Gerber, Ph, Bacroix, B.: Estimation of recrystallized volume fraction from EBSD data. Acta Mater. **50**, 1467–1477 (2002)
8. Wu, G., Jensen, D.J.: Automatic determination of recrystallization parameters based on EBSD mapping. Mater. Char. **59**, 794–800 (2008)
9. Humphreys, F.J.: Grain and subgrain characterisation by electron backscatter diffraction. J. Mater. Sci. **36**, 3833–3854 (2001)
10. Larsen, A.W., Jensen, D.J.: Automatic determination of recrystallization parameters in metals by electron backscatter pattern line scans. Mater. Char. **51**, 271–282 (2003)
11. Benchabanea, G., Boumerzouga, Z., Thibonb, I., Gloriant, T.: Recrystallization of pure copper investigated by calorimetry and microhardness. Mater. Char. **59**, 1425–1428 (2008)

# Aviation Safety Aspects of the Use of Unmanned Aerial Vehicles (UAV)

Bertold Békési, Mátyás Palik, Tímea Vas
and Alexandra Halászné Tóth

**Abstract** During recent years Unmanned Aerial Vehicles (UAV's) have been used in more and more versatile ways, both in the military and civilian life. Their use in certain fields of aviation is expanding rapidly. At present practically all organizations, authorities and working groups involved in aviation safety and flight safety consider this specific set of instruments technology study it's regulatory and safety aspects and issues related to its use. The aim of our study group, within the consortium project, was a comprehensive research into the conditions of the national use of unmanned aerial vehicles from the perspective of air traffic safety.

**Keywords** UAV · UAS · Unmanned aerial vehicles · Unmanned aircraft systems · Drone · Air traffic · Air law · Safety

## 1 Initial UAV Developments and Results of Research Carried Out in Hungary

The beginning of UAV history in Hungary is marked by the development project code named Szojka III. The multipurpose, small unmanned aircraft system was developed as part of the defense industry cooperation with Czechoslovakia in 1988.

B. Békési · M. Palik (✉) · T. Vas
Institute of Military Aviation, National University of Public Services, Szolnok, Hungary
e-mail: palik.matyas@uni-nke.hu

B. Békési
e-mail: bekesi.bertold@uni-nke.hu

T. Vas
e-mail: vas.timea@uni-nke.hu

A.H. Tóth
Office of Authorities Military Aviation Directorate, Ministry of Defense, Budapest, Hungary
e-mail: toth.alexandra@hm.gov.hu

© Springer International Publishing Switzerland 2016
L. Nádai and J. Padányi (eds.), *Critical Infrastructure Protection Research*,
Topics in Intelligent Engineering and Informatics 12,
DOI 10.1007/978-3-319-28091-2_10

In that project the Czech party supplied the airframe and servos for flight control. Onboard and ground elements of the control system, navigation equipment, ground control station and payload were developed by the Hungarian party. The aircraft's adverse maneuverability characteristics caused severe difficulty for the project, landings often ended up with aircraft crash. The project was terminated on mutual consent and assets were divided between the developers. Subsequently, the Czech armed forces went on with the development of its own complex, and had it in service till December 2011.

Since 1999 AeroTarget Limited Partnership and its legal predecessor, AeroMeat Ltd. have been developing, operating and manufacturing the family of Meteor-3 aerial target drone for the Hungarian Defense Force. These aircraft are mainly used for ground based air defense personnel training during missile firing exercises. In 2005 the company with several years of experience in manufacturing and operation won the tender for the Hungarian Defense Force's aerial target aircraft modernization.

Using experience gained, new principles of UAV trajectory design were introduced, and a new set of updated specifications and requirements were developed for these vehicles. Meteor3M then were developed further, and the jet engine Meteor-3MA military aerial target aircraft (Fig. 1) has become a big Hungarian success; it took two years to take this aircraft from the design phase to field testing [1].

In 2007 the HM EI Zrt. (MOD ED Co.) started to develop drones capable to carry out reconnaissance missions as well, addressing the needs of the Hungarian Defense Force. Members of this vehicle family were introduced to public in autumn 2012. In the course of the creation of reconnaissance UAV's the experts of the company made good use of the knowledge gained through the development of the previous types of aerial target aircraft. The aircraft and its server units were produced exclusively with the work of Hungarian experts, using Hungarian raw materials and software [2].

The consortium established in 2008 under the leadership of BHE Bonn Hungary Elektronikai Kft. (BHE Bonn Hungary Electronics Ltd.) with the participation of Budapest University of Technology and Economics Mobile Innovation Center and

**Fig. 1** The newly developed Meteor–3MA TUAV

Óbuda University John von Neumann Faculty of Informatics aimed at the development of a UAS for civilian purposes (Fig. 2).

The development resulted in the prototype of a cost effective UAS belonging to the mini category, which does not require any specific infrastructure and can be used extensively for civilian, disaster relief and search and rescue missions. The highly sophisticated onboard and central control intelligence provide for the construction's fully automated operation. Due to this fact the range of possible users may be much wider than that of other, similar aircraft [3].

In the recent decades several study groups of the Hungarian military higher education establishments together with several developers and manufacturers have been involved in intense R&D efforts regarding the issues of the use of the UAV's. These workshops were set up with the objective to address 21st century challenges the militaries have to face. The researchers consider that their mission is to determine directions of development, production and training of the personnel, which are the priorities for higher education, scientific research and technological modernization. Using the cascading method in their R&D work and with appropriate concentration of material resources the participants' aim was to provide full scope solutions for the use of UAV's in Hungary. Their efforts mainly focus on the research and development of UAV airframes, engine, onboard sensors, as well as payload, control and navigation equipment and autopilots. Their research was extended to the development of the requirements and regulations necessary for the use and operation of UAV's in military and civilian missions. Theoretical research was followed by test flights where payload usability was also tested. Research and test results were continuously recorded; conclusions were drawn, and respectively used in further phases. As a result of R&D efforts, some 15 fixed wing and rotary wing UAV prototypes have been produced. Several participants of the project have achieved considerable professional success in the implementation phase [4].

Another Hungarian development is related to this branch of research: academics and companies participating in the project have a major role in the event that happened in the small Hungarian town of Szendrő on 14 August 2006, when the first in the world small UAV, was used for fire reconnaissance purposes by the local fire department [5].

**Fig. 2** BHE demonstration at the 2013 conference on unmanned aerial systems

R&D efforts are ongoing, more and more specialists, young scientists, academics and students get involved in the developments. In the relevant university, faculty and department communities there have been produced several PhD dissertations, thesis papers and papers for the scientific students' associations' conferences. Now the industrial background mentioned above is capable of producing marketable unmanned aerial vehicles. There is a huge demand both from the defense sector and the civilian sector, but the supply is big as well, especially in the international arena. That is the reason why only exceptionally reliable and high value added products can be sold.

## 2  Limitations of the Use of UAV's

What makes it a topical issue to study the use of UAV's from air traffic safety perspective and to develop its regulatory background is that current regulations in force are incomplete; the legal framework for the domestic use of UAV's is missing. Earlier in Hungary the topic had not been covered in a comprehensive way, either from the perspective of the use, or the perspective of air traffic safety or the adopting environment.

Considering this fact, the main objectives of our research are to study the conditions for the current and future use of UAV's, and to process international recommendations already available, as well as determine regulatory framework and make recommendations for the governmental organizations involved in decision making and execution, and for the customers regarding the modification of the legal background, and respectively regarding its appropriate design; furthermore, to make suggestions to create technical conditions and conditions of use for UAV's.

## 3  Research Results

After the acceptance of the tender, detailed research and publication plans and financial perspectives were developed, and tasks and deadlines for individual researchers and study groups were determined. Then we collected effective national legislation documents, recommendations and documents of international aviation, air navigation and air traffic safety organizations.

Then we analyzed and compared documents focusing on the designation of the areas which lack in regulation, and areas representing significant risk to air traffic safety. We pointed out the shortcomings in existing legislation, applied technological equipment and operating conditions; in addition, the impact of those shortcomings on air traffic safety was analyzed. Then the identical features of different regulations were identified; the advantages of existing regulatory and technological systems were highlighted, and possibilities for their further development were also described.

To provide for a coherent approach, we developed a generally acceptable system of terminology and abbreviations/acronyms, which has become a common starting element that is an inevitable condition for the establishment of the future regulatory environment.

In the further phases of our study we made recommendations for the development of the UAV legislative framework. Our study group prepared the study "Impact of Current Legislative, Technical and Operational Criteria of Unmanned Aerial Vehicles on Air Traffic Safety" based on the legislation concerning international and national air traffic. In this paper we studied our national legislative documents of different levels, which address several areas of air traffic including the issue of mandatory liability insurance as well [6]. The paper on the conduct of the investigation highlights those provisions of individual legislations which are questionable concerning the use of UAV's or require new regulation. In some cases suggestions were made regarding the modification or supplementation of current legislation [7, 8]. The planned rule making process is shown in Fig. 3.

The study group dealing with technological issues studied the issues of UAV categorization, requirements for their technical equipment and their onboard hardware elements. This study resulted in a number of research reports, papers and presentations at national and international conferences [9–13]. Research findings in this field are summarized in the study "UAV Categorization, Onboard Hardware Classification".

The researchers were expected to determine the set of technological and technical instruments of the UAV's guaranteeing air traffic safety, and then they had to make recommendations regarding onboard equipment. Finally, the study "The Set of UAV's Technological and Technical Instruments" analyzes operation, servicing and technical procedures [14–16].

In order to achieve success in research, it is inevitable to approach the subject of the research with scientific thoroughness. To meet this requirement, in addition to the methods of analysis and synthesis we used other techniques as well.

For statistical analyses we carried out data collection using online questionnaires, through which useful information was gathered from potential users (national defense, disaster defense, law and order organizations, environmental protection organizations, transportation, industry, agriculture, media, etc.) regarding their concepts of the use of UAV's.

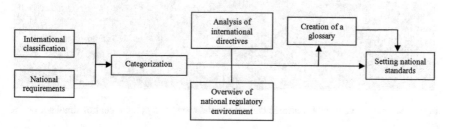

**Fig. 3** Planned rulemaking process

To investigate the operation of UAV's from military aerodromes, we carried out flight safety conflict analysis based on the following EUROCONTROL publications: EUROCONTROL Safety Regulatory Requirement 4 and Air Navigation System Safety Assessment Methodology 2.1. In the investigation there were involved air traffic controllers, civilian and military pilots, UAV operators, air traffic advisors and flight safety experts. Within the scope of the analysis exclusively the basic conditions of UAV integration into the relevant air space, as well as the risks of integration, and risk management were studied; issues related to take off from the aerodrome and landing on the aerodrome were not investigated.

To support our hypotheses, simulation exercises were carried out in 3D aerodrome and terminal radar control simulators. To make it more true to life we had departure, arrival and holding procedures planned, and prepared exercises mirroring real world situations.

In the aerodrome environment investigations were conducted in the 3D tower simulator (which is at an International level) of HungaroControl, Hungarian Air Navigation Services Pte. Ltd. Co., while simulation of the traffic in the terminal environment was carried out in the LETVIS radar simulator of the Military Aviation Department of the National University of Public Service (Fig. 4). One of the final reports describes the specific features of the application of UAV, while the other one presents main features of air traffic controllers' workload and changes in their workload. Both series of exercises were conducted during five days each, with five or six simulations each day lasting 45 min as a maximum. In the execution of the simulations highly experienced, licensed air traffic controllers were involved [17].

In order to provide for the safe UAV operations both in the air and on the ground, it is inevitable to develop recommendations regarding general and specific flight rules. The summary paper "General and Specific Flight Rules of UAV Flights" was completed in the second phase of the project. This paper contains specific recommendations for the conduct of UAV flights and requirements determined by the aviation authority; it also deals with the issues of aerodrome operation and the set of flight related operational documents [16, 18, 19]

**Fig. 4** Simulation venues. Aerodrome control simulator on the *left*, radar control simulator on the *right*

Our last study paper is "Supporting and Organizational Requirements of UAV flights", which contains recommendations regarding flight planning and management, as well as the issues of the use of the air space [20–22].

Apart from the papers published earlier, research reports and studies reviewed above, one of the main indicators of the tender is the summary study "UAS Strategy—National Use of Unmanned Aerial Systems". The document describes the main results of the research conducted, as well as the most important recommendations for the legislators and other governmental organizations. The Authors present their visions concerning the future use of UAS; describe their characteristics, classification principles, possibilities of their use (military, public and other) at present. Authors also identify regulatory and technical criteria and expectations (regarding technical minima for onboard systems, technical design, operational documentation and safe operation). Among organizational issues, UAV operations, training and R&D were studied.

One of the objectives of the project was to promote unmanned aviation and familiarization with it. To meet this objective the researchers compiled a book disseminating scientific knowledge "Unmanned Aviation for Professionals and Amateurs" [23]. The book provides useful information for amateurs interested in the subject, and for professionals having more thorough knowledge in aviation, too.

In this field of the research more than 10 people, 50 % of them lecturers-researchers with PhD participated with permanent commission, and about 20 other experts who wrote more than 60 national and international papers, 11 research reports, 5 studies and one strategy. During the two year project participants gave more than 50 presentations at conferences in Hungary and 6 other countries. Several BsC, MsC and PhD students of our university also joined our study group. They published their research results at conferences, in thesis papers and papers made for scientific students' associations conferences.

We consider research results to be useful in the following main areas:

– Governmental organizations involved in rule making (Ministry of Defense, Ministry of Interior, Ministry of Public Administration and Justice, Ministry for National Development and Economy, National Air Space Coordination Working Group, etc.)
– Organizations involved in conducts of flights (Hungarian Defense Force, airlines, aviation associations, flying clubs, training institutions, etc.)
– All participants of air traffic management (HungaroControl, Hungarian Air Navigation Services Pte. Ltd. Co., military air defense and air traffic control units, etc.)
– All participants of aerodrome operation (Budapest Airport, Hungarian Defense Force and civilian operators respectively)
– All present and future organizations using and operating UAV

Success gained in the focal research area can promote the expansion of the use of UAV's in Hungary and the enhancement of air traffic safety; it increases competency of researchers (professionals) involved in the project. Above all, the project contributes to the implementation of (long term objectives and strategic goals of the participating organizations.

There are opportunities for the implementation of long term objectives of the research after the project is finished. Efforts invested into research can be considered a success if its results are implemented, that is, if and when the documents regulating all aspects of national UAV flying are developed.

**Acknowledgements** We acknowledge the financial support of this work by the Hungarian State and the European Union under the TÁMOP-4.2.1B-11/2/KMR-2011-0001 project.

# References

1. Kovács, L., Ványa, L.: Unmanned aerial vehicles in the Hungarian defense forces. In: 10th International Symposium of Hungarian Researchers on Computational Intelligence and Informatics, Budapest, pp. 753–763 (2009)
2. Méhes, L.: Running unmanned aerial vehicles in Hungary. In: International Conference of Scientific Paper AFASES 2013, Henri Coanda Air Force Academy, Brasov, pp. 646–650 (2013)
3. Mikó, G., Kazi, K., Solymosi, J., Földes, J.: UAV development at BHE Bonn Hungary Ltd. In: 10th International Symposium of Hungarian Researchers on Computational Intelligence and Informatics, Budapest, pp. 803–820 (2009)
4. Makkay, I.: Robotics in the 21st century. Goals and some results of science research works at national defense university. In: Academic and Applied Research in Military Science, vol. 2, no. 2, pp. 175–184. ZMNDU, Budapest (2003)
5. Restás, Á.: Robot reconnaissance aircraft for fighting forest fires. In: Academic and Applied Research in Military Science, vol. 3, no. 5, pp. 653–664. ZMNDU, Budapest (2004)
6. Tóth, A.: Reflections on the aviation liability insurance of state UAV with a perspective on critical infrastructure protection. In: Management—Theory, Education and Practise 2013, pp. 119–123. Armed Forces Academy of General M. R. Stefanik, LiptoV'sky Mikulás (2013)
7. Tóth, A., Somosi, V., Pongrácz, G.: The current issues of defining basic concepts for the national regulations related to the unmanned aerial vehicles. In: Deterioration, Dependability, Diagnostics, pp. 155–160. University of Defense Faculty of Military Technology Department of Combat and Special Vehicles, Brno (2012)
8. Tóth, A., Somosi, V.: Risk factors of the civil use of UAV's. In: Proceedings of 16th International Conference. Transport Means 2012, Technologija, Kaunas, pp. 211–213 (2012)
9. Simon, S.: Concept for UAV's maintenance publication and operation manual system; UAV categorization. In: Deterioration, Dependability, Diagnostics, pp. 149–158. University of Defense Faculty of Military Technology Department of Combat and Special Vehicles, Brno (2013)
10. Makkay, I.: Redundancy for UAS—Failsafe engineering. In: Proceedings of 16th International Conference. Transport Means 2012, Technologija, Kaunas, pp. 192–195 (2012)
11. Békési, B.: Redundancy on board of UAV's—energy systems. In: Proceedings of 16th International Conference. Transport Means 2012, Technologija, Kaunas, pp. 158–161 (2012)
12. Bali, T., Palik, M.: Tactical UAV onboard systems. In: Management—Theory, Education and Practise 2013, pp. 24–31. Armed Forces Academy of General M. R. Stefanik, LiptoV'sky Mikulás (2013)
13. Szabolcsi, R.: Conceptual design of the unmanned aerial vehicle systems used for military applications. In: Scientific Bulletin of Henri Coanda Air Force Academy, no. 1, pp. 61–68. Henri Coanda Air Force Academy, Brasov (2009)

14. Makkay, I.: Redundancy for UAV's—ground control stations. In: Bulletins in Aeronautical Sciences, vol. 25, no. 4, pp. 46–52. NUPS, Szolnok (2013)
15. Wührl, T.: Redundancy for micro UAV's—onboard control system redundancy. In: Proceedings of 16th International Conference. Transport Means 2012, Technologija, Kaunas, pp. 62–64 (2012)
16. Ozoli, Z.: Type certification and airworthiness certification procedure of unmanned aircraft. In: International Conference of Scientific Paper AFASES 2013, Henri Coanda Air Force Academy, Brasov, pp. 670–673 (2013)
17. Fekete, C.: UAV simulations in the protection of critical infrastructure research. In: Management—Theory, Education and Practise 2013, pp. 75–79. Armed Forces Academy of General M. R. Stefanik, LiptoV'sky Mikulás (2013)
18. Simon, S.: Design and production criteria and authority requirements for the hungarian defense forces operated UAS. In: International Conference of Scientific Paper AFASES 2013, Henri Coanda Air Force Academy, Brasov, pp. 691–695 (2013)
19. Makkay, I.: Beyond visual line of sight—FPV flight. In: New Trends in Civil Aviation 2013, pp. 58–69. The Faculty of Operation and Economics of Transport and Communications, Air Transport Department and Air Training and Education Center, Zilina (2013)
20. Vas, T., Palik, M.: UAV operation in aerodrome safety and ACS procedures. In: The 7th International Scientific Conference Defense Resources Management in the 21st Century, National Defense University, "Carol" I Publishing House, Brasov, pp. 75–83 (2012)
21. Bottyán, Z., Wantuch, F., Gyöngyösi, A., Tuba, Z., Hadobács, K., Kardos, P., Kurunczi, R.: Development of a complex meteorological support system for UAV's, World Academy of Science, Engineering and Technology, vol. 7, no. 4, pp. 648–653 (2013)
22. Pongácz, G., Palik, M.: Communication issues of UAV integration into non segregated airspace. In: The 7th International Scientific Conference Defense Resources Management in the 21st Century, National Defense University, Carol I Publishing House, Brasov, pp. 69–74 (2012)
23. Békési, B., Bottyán, Z., Dunai, P., Tóth, A., Makkay, I., Palik, M., Restás, Á., Wührl, T.: Pilóta nélküli repülés profiknak és amatőröknek. Nemzeti Közszolgálati Egyetem, Budapest (2013)

# Cloud Security Monitoring and Vulnerability Management

M. Kozlovszky

**Abstract** Cloud computing security is a fundamental concern. One of the key problems is how one can test, monitor or measure the underlying Cloud infrastructure from user/customer space. Our aim is to build up tools and solutions to measure and assess quantitative and qualitative security parameter values of a generic IaaS cloud system. We have created a measurement framework (Cloudscope), which is capable to measure the targeted IaaS cloud system from security point-of-view automatically. Furthermore, we have built an easy-to-extend framework to assess the examined cloud infrastructure. Our solution can be used by potential tenants/end-users and governmental organizations to evaluate and assess IaaS type cloud systems. In this paper we present our virtualized cloud security monitor and assessment solution, we describe its main functionalities.

**Keywords** Cyber defense · Cloud assessment · Information security

# 1 Introduction

## 1.1 Aim

In recent years, the adoption of new technologies (such as unified communications, server virtualization, high-speed networking and cloud services) created a difficult task for operational teams responsible for managing and securing corporate networks and data centers. Instead of just diagnosing and troubleshooting problems now they need to cope with the communication over increased bandwidth and protection of fully virtualized data centers.

M. Kozlovszky (✉)
Óbuda University/John von Neumann Faculty of Informatics, Budapest, Hungary
e-mail: kozlovszky.miklos@nik.uni-obuda.hu

© Springer International Publishing Switzerland 2016
L. Nádai and J. Padányi (eds.), *Critical Infrastructure Protection Research*,
Topics in Intelligent Engineering and Informatics 12,
DOI 10.1007/978-3-319-28091-2_11

From the literature we can identify four basic cloud deployment models: public, private, community and hybrid. In all these models, there is a strong need to know precisely how the services and the underlying infrastructure are performing and how security of the whole system. In three, from the four models (public, community and hybrid cloud environments), a huge information and knowledge gap exists between the service provider and the customers. In an ideal world all key players of the cloud ecosystems are able to track what is going on in the system at their level. Cloud acceptance is highly correlated as too directly how transparent and reliable the cloud services are offered by the service providers. In multitenant clouds, the customer should fully trust the cloud service provider (how the provider protects its hardware infrastructure, how secure the premises are, how low the oversubscription factor is, or how reliable the communication channels are, etc.). Moreover, if the customer should be compliant with some business domain specific standards or regulations, the compliance should be validated somehow for the cloud service provider's infrastructure/services.

The complexity of a generic IaaS type cloud infrastructure with hosted multi-tenant, virtualized business environments is definitely high (generic cloud example shown in Fig. 1).

Additionally, nowadays a few vendor specific and open source middleware solutions are dominating the cloud market. Commercial cloud middleware solutions are offered by a few big players (Vmware, Cisco, HP, IBM, and Amazon) and a vast amount of smaller enterprises. Major competitors in open source IaaS cloud branch are Eucalyptus [1], Nimbus/Cumulus [2], Open Nebula [3] and OpenStack [4]. Cloud adopters are using cloud technologies at different maturity to gain benefits and they can only hope and use non-standardized, non-transparent, un-trustable infrastructure and services. Clouds are frequently used for business purposes, so end user's data protection is a more important topic in clouds than in

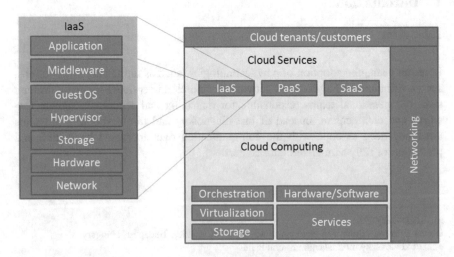

**Fig. 1** Generic IaaS type cloud infrastructure environment

any other Distributed Computing Infrastructure (DCI's). Our general goal is to provide assistance for some of the key players such as system administrators building up sustainable and less vulnerable infrastructure and survive cyber-attacks. For this aim we have created an IT security audit framework for distributed computing systems (DCI's) in order to assess, evaluate and decrease vulnerability level of the underlying cloud infrastructure.

In Sect. 1 we shortly introduce recent vulnerability trends. In Sect. 2 we provide information about existing generic and security focused DCI monitoring solutions. In Sect. 3 we describe the design of the Cloudscope modular cloud evaluation and monitoring framework and show the security module of the solution with various integrated Vulnerability Assessment Tools. In Sect. 4 we detail cloud evaluation basics. In Sect. 5 we give an overview about our cloud assessment solution with the description of our assessment procedure workflow. At the end of our paper, we conclude and explain the direction of our future work.

## 1.2 Vulnerability Trends

The software evolution (e.g. size, functionality set) generally increases complexity in the software stack. It is hard to protect even a single PC node infrastructure against malicious attacks. This problem multiplies significantly if the infrastructure is heavily distributed, contains thousands of cores, and serves hundreds of people. During recent years the amount of explored network vulnerabilities are increasing constantly with about 500 Network Vulnerability Tests (NVT's) per month [5].

Cloud computing technology is rapidly spreading within the IT world. Companies worldwide migrate their infrastructure to cloud infrastructure. However there are lots of known potential security problems with multi-tenant, always available, online, distributed and shared systems:

- Caused by the technology and the complexity itself
- Caused by site/software stack setup
- Caused by end-user behavior

Due to the massively virtualized, up-scaled environment and the homogeneity of the infrastructure a successful break-in method is very likely become instantly applicable against other similar cloud environments. Usually it takes non-zero time to find the vulnerability, initiate alarm and document protection by the incident response teams. As a consequence a valid vulnerability can be likely reusable and provide a vast gain for the potential intruders on a short term basis.

## 2 Cloud Benchmarking and Monitoring

Benchmarking is by definition the process of comparing one's business processes and performance metrics to industry best, to enable comparisons to be made between providers, in an objective, reproducible manner. Dimensions typically measured are quality, time and cost. Benchmarking is used to measure performance using specific indicators (cost per unit of measure, productivity per unit of measure, cycle time of x per unit of measure or defects per unit of measure) resulting in a metric of performance that is then compared to others [6].

Distributed Computing Infrastructures (DCI's), such as, Clouds, Supercomputers or Grids, are monitored for various reasons and by various methods. Widely used monitoring categories are the followings: status monitoring, performance monitoring and security monitoring: Monitoring can be realized both by active and passive-software/hardware-elements. Monitoring software solutions can be native/de facto standard applications (e.g. SNMP based status monitoring, Linpack based performance monitoring, pattern based security monitoring, NVT/CVE based vulnerability scanning [7, 8], patched/modified software solutions, or self developed monitoring ones. For full-scale cloud monitoring, administrators are keen to combine existing and in-house developed software solutions together. In the following we will focus only on some of the available generic cloud benchmarking and vulnerability/security monitoring solutions.

## 2.1   Amazon CloudWatch

Amazon CloudWatch provides monitoring for AWS cloud resources and the applications customers run on AWS. Developers and system administrators can use it to collect and track metrics, gain insight, and react immediately to keep their applications and businesses running smoothly. Amazon CloudWatch monitors AWS resources such as Amazon EC2 and Amazon RDS DB instances, and can also monitor custom metrics generated by a customer's applications and services. With Amazon CloudWatch, system-wide visibility into resource utilization, application performance, and operational health can be gained [9]. For Amazon EC2 instances, Amazon CloudWatch Basic Monitoring collects and reports metrics for CPU utilization, data transfer, and disk usage activity from each Amazon EC2 instance at a five-minute frequency. Amazon CloudWatch Detailed Monitoring provides these same metrics at one-minute intervals, and also enables data aggregation by Amazon EC2 AMI ID and instance type. For Auto Scaling or Elastic Load Balancing, Amazon CloudWatch will also provide Amazon EC2 instance metrics aggregated by Auto Scaling group and by Elastic Load Balancer. Monitoring data is retained for two weeks this enables browsing historical monitoring data.

## 2.2   Grid Site Software Vulnerability Analyzer

Grid Site Software Vulnerability Analyzer (GSSVA) [10] by MTA SZTAKI is a monitoring tool which collects status information of the distributed computing infrastructure machines, analyzes the information gathered and compares the results

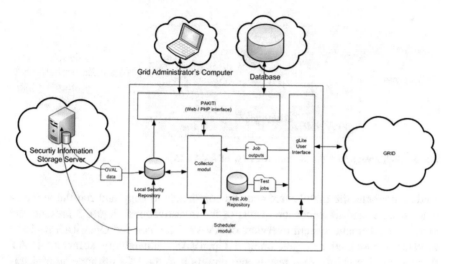

**Fig. 2** High level schematic system overview of GSSVA [10]

using an external information repository to find the existing security problems. It can automatically explore the installed Linux packages of the computing elements (CE) and worker nodes (WN). It is using a modified status monitoring system as basis software called PAKITI [11]. PAKITI, is basically a patching status monitoring tool, which can be used for infrastructure security status monitoring. GSSVA collects the list of the installed software packages on the nodes and matches the gathered information with the security database (coming from an external repository). Figure 2 shows the high level schematic system overview of GSSVA. It is using HTTP or HTTPS protocol to communicate and provide a graphical user interface for its users. GSSVA is an official IT security software tool of SEE-GRID-SCI project's infrastructure monitoring solution.

### 2.3 OpenVAS

OpenVAS—Open Vulnerability Assessment System [12] is an open source vulnerability scanner and vulnerability management solution. Beside that it is also a framework, which contains several services, tools and a continuously increasing number of (about 30,000) Network Vulnerability Tests (NVT's).

### 2.4 Advanced Vulnerability Assessment Tool

Advanced Vulnerability Assessment Tool (AVAT) [5] by MTA SZTAKI supports various DCIs (e.g. grids such as ARC and gLite and clouds). The DCI interface

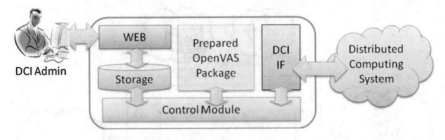

**Fig. 3** High level schematic system overview of AVAT [5]

module contains the middleware specific commands to copy and run the vulnerability scanner and to gather the results of the investigations. Figure 3 presents the high level schematic system overview of AVAT. The included OpenVAS package contains a modified and precompiled OpenVAS vulnerability scanner. AVAT stores the vulnerability scan results and reports it to the DCI resource administrators. AVAT is used within the HP-SEE project to assess vulnerability on its available supercomputing infrastructure.

## 2.5 Nessus

Nessus [13] by Tenable Network Security Inc. provides centralized management of multiple vulnerability scanners and real-time vulnerability, log, and compliance management. Since 2005 the software is not open source anymore. OpenVAS is its follow up open source project.

## 2.6 Nexpose and Metasploit

Nexpose [14] by Rapid7 Inc. provides a full scale (from single user, up to enterprise level) vulnerability management solution. It was among the fist software solutions, which received USGCB (United States Government Configuration Baseline, and EAL3+ certificates Common Criteria Certification for Evaluation Assurance level Augmented [15]. It supports automatic asset discovery, scanning and remediation on virtualized environments. Metasploit [16] is an open source vulnerability scanner solution, focusing mainly on penetration testing. Recently the two software solutions are combined closely together. Metasploit is a handy solution to validate security risks, audit IT infrastructure and verify vulnerabilities with simulated penetration tests.

## 2.7  Microsoft Baseline Security Analyzer

The Microsoft Baseline Security Analyzer (MBSA) [17] is a freeware tool designed to determine security state most of the MS based environment in accordance with Microsoft security recommendations and offers specific remediation guidance. It is a standalone security and vulnerability scanner designed to identify common security misconfigurations and missing security updates. Its availability and importance decreased during recent years.

## 2.8  Qualys Guard

Qualys Guard [18] by Qualys Inc. helps organizations with globally distributed data centers and IT infrastructures to identify their IT assets, collect and analyze large amounts of IT security data, discover and prioritize vulnerabilities, recommend remediation actions and verify the implementation of such actions.

## 2.9  NetIQ Security Manager

NetIQ Security Manager by NetIQ [19] is a Security Information and Event Management (SIEM) solution that provides host-focused security, supports automatic security activity reviews, log collection, threat management, incident response, and change detection. It is capable to do change detection and file integrity monitoring, privileged-user monitoring, log management, analysis and query-based forensics.

## 3  Combined Infrastructure Vulnerability Scanning— Cloudscope

Centralized vulnerability assessment solutions for distributed systems are only focusing on some fragments of the whole IT security picture and feature-rich frameworks (e.g. OpenVAS, Metasploit) could not work easily in multi-administration domains frequently used in DCIs. To overcome these issues we designed our IT software security solution—Cloudscope—to combine the benefits of various IT security software tools and frameworks. Moreover, we combined existing vulnerability scanner and monitoring tools (such as GSSVA used on girds/EGI and SEE-GRID-SCI/, and AVAT used on HP-SEE supercomputing infrastructure and on private clouds) and tried to use these solutions in an integrated way.

We have developed a modular software tool, which contains, as pluggable modules, various software solutions and able to examine the infrastructure, from a security point-of-view. We offer a clear, adaptable, concise and easy-to-extend framework to assess the underlying DCI infrastructure. Our developed solution is generic and multipurpose it can act as a vulnerability scanner and has support for asset discovery at the same time. It is virtualized and collects assessment information by the decentralized Security Monitor and it archives the results received from the components and visualize them via a web interface for the testers/administrators. In our Vulnerability Management module we are able to measure vulnerability parameters of a certain DCI. We are using parallel the integrated security measurement applications.

## 3.1  Vulnerability Assessment

Our solution collects information about security status using various external software tools, and matches the gathered information with security data. It is using HTTP and HTTPS protocols to communicate and provide security information via the graphical user interface. The framework is capable to test various virtualized and non-virtualized DCI's (such as grids, clouds, normal/HPC clusters), and due to its own virtualized environment it can be used as a generic security assessment tool to evaluate even non-virtualized critical infrastructure.

## 3.2  Internal Architecture

Cloudscope is trying to remain middleware, virtualization technology and OS independent. The vulnerability testing module orchestrate remotely the loosely coupled various scanning tools. These tools are gathering the information and store their results internally.

All integrated virtualized security scanner software works in an agent-like way, separately. Both the client side (which contains the vulnerability/security probes and scanners) and the server size (which contains the local data collectors of each individual probes and scanners) are virtualized. Each plug-in is controlled remotely.

The user can initiate a vulnerability scanning task on the web front-end (shown in Fig. 4). Firstly an XML based work list is generated according to the user's security scanning aim. This work list with a large set of configuration parameters

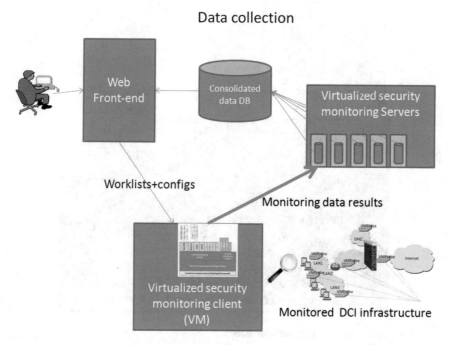

**Fig. 4** Cloudscope, combined infrastructure vulnerability scanner (schematic overview)

forwarded to the virtualized security monitoring client automatically. Each pre-defined work task of the work list runs separately on the VM and the vulnerability scanners examining the infrastructure individually.

The results are forwarded automatically by the scanners to their own virtualized servers. Cloudscope extracts the relevant security results and stores into a consolidated database. This consolidated database is connected together with the centralized web front-end, where the user can see the annotated data of the vulnerability scanners. Our solution is capable to check not only single spot problems, but able to examine multi-point vulnerability tests if needed. The internal architecture of the data collection framework is modular and enables easy integration of all kind of measurement tools (shown in Fig. 5).

We post-process and evaluate the collected data on the centralized server side of the framework and provide feedback on the web user interface about the examined system. Inside our developed security assessment framework standard DB (MySQL) and Web server (Apache based) solutions have been used. The framework is open source, and there are no restrictions to re-implement and/or adapt it to other community needs. Dependencies on proprietary or commercial products are optional, due to the loosely coupled external tools.

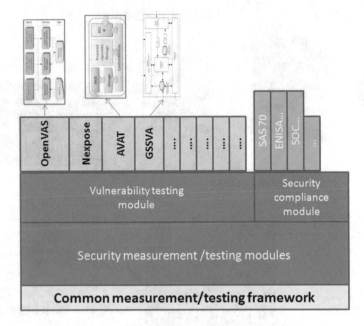

**Fig. 5** Internal module structure of Cloudscope

# 4    Cloud Evaluation

## 4.1    Existing Best Practices

Cyber security is one of the major driving forces behind cloud infrastructure and
service evaluation. IT security organizations and groups such as CSA [21] (Cloud
Security Alliance), ENISA [22] (European Network and Information Security
Agency), the Cloud Computing Interoperability Group, and the Jericho Forum [23]
are actively doing cloud ecosystem evaluation mainly from security, and data
control point of view. The U.S. government initiated the FedRAMP [24] (Federal
Risk and Authorization Management Program, which is basically a risk management program for large outsourced and multi-agency information systems. It
defines the entire assurance process for cloud instances, provides compliance
evaluation for individual governmental agency applications and also authorizes
federal IT services. There are worldwide de facto audit standards of enterprise
financial and infrastructure-related internal controls: such as the Statement on
Auditing Standards No. 70 report usually referred to as SAS 70 [25], the SSAE 16
[26] and SOC reports [27] and the CSA guide [28].

## 4.2 IaaS—Type Cloud Evaluation

Information as a Service (IaaS) type cloud infrastructure has a complex structure. To evaluate cloud providers with their IaaS type infrastructure one needs to identify a large set of complex evaluation parameters (criteria). All the parameters should fulfill the following basic requirements:

- Parameter/criteria should be objective
- Parameters should be measurable
- Parameter should be relevant from cloud user's and/or service provider's point of view

### 4.2.1 Evaluation Categories

We have identified 19 main evaluation parameter groups (shown in Table 1).

These groups provide us an exhaustive parameter field for Cloud Service Provider assessment [20]. The evaluation parameter set has a tree shape graph

**Table 1** Parameter categories

| No | Parameter group | Notes |
|----|-----------------|-------|
| 1 | Generic parameters | Generic cloud provider parameters (size, facilities…) |
| 2 | Generic functionalities | Generic cloud provider services and functionalities |
| 3 | Performance (infrastructure) | Cloud provider's infrastructure performance |
| 4 | Quality (infrastructure) | Cloud provider's infrastructure quality (homogeneity) |
| 5 | Scalability | Cloud provider's infrastructure sea lability |
| 6 | Security | Infrastructure and service security as a whole |
| 7 | High availability/disaster recovery | Cloud provider's infrastructure stability, error proneness |
| 8 | Viability | Cloud provider's business stability, survivability |
| 9 | QoS | Cloud provider's service and infrastructure quality, and their assurance |
| 10 | Standards | Cloud provider's (re)used standards in services and infrastructure |
| 11 | Monitoring | Cloud provider's monitoring capabilities |
| 12 | Accounting/logging | Cloud provider's accounting/logging capabilities |
| 13 | Data maintenance | Cloud provider's data maintenance capabilities |
| 14 | User friendlyness | Cloud provider's service and infrastructure usability |
| 15 | Green/efficiency | Cloud provider's energy efficiency parameters |
| 16 | Transparency | Cloud provider's business/service transparency |
| 17 | Legal and regulatory compliance | Cloud provider's legal and regulatory compliances |
| 18 | Service and infrastructure assessment | Cloud provider's existing service and infrastructure assessments |
| 19 | Knowledge base and competence | Cloud provider's competence level in its services and infrastructure |

**Table 2** Security assessment parameter sub-groups

| No | Parameter sub-groups | Notes |
|------|------------------------------------|----------------------------------------------------------|
| 1 | IT human resources | IT personnel (size, etc.) |
| 1.1 | Dedicated security personnel | IT security personnel (certified, non-certified, etc.) |
| 2 | Incident management | Incident management parameters |
| 2.1 | IT security incident management | IT incident response team parameters |
| 2.2 | Business incident management | Business incident response team parameters |
| 3 | Risk management | Risk management parameters |
| 3.1 | Asset enumeration | Asset enumeration parameters |
| 3.2 | Risk enumeration | Risk enumeration parameters |
| 3.3 | Business impact evaluation | Business impact parameters |
| 3.4 | Reevaluation | Reevaluation parameters (periodical, duration, etc.) |
| 4 | Vulnerability management | Vulnerability management parameters |
| 4.1 | Infrastructure vulnerabilities | Infrastructure vulnerability parameters (stats, etc.) |
| 4.2 | Application vulnerabilities | Application vulnerabilities parameters (stats, etc.) |
| 4.3 | Service vulnerabilities | Service vulnerabilities parameters (stats, etc.) |
| 4.4 | Accessed vulnerability repositories | Vulnerability repository parameters |
| 4.5 | Automatic vulnerability alarming | Automatic vulnerability alarming parameters |
| 4.6 | Automatic vulnerability monitoring | Automatic vulnerability monitoring parameters |
| 4.7 | Automatic vulnerability elimination | Automatic vulnerability elimination parameters |
| 5 | Identity management | Identity management parameters |
| 6 | Patch management | Patch management parameters |
| 7 | Authentication | Authentication parameters |
| 8 | Authorisation | Authorisation parameters |
| 9 | Accounting | Accounting parameters |
| 10 | Infection management | Infection management parameters |
| 11 | Encryption | Encryption parameters |
| 12 | Access management | Access management parameters |
| 12.1 | Physical access contrail | Pysical access contrail and prevention parameters |
| 12.2 | Digital access contrail | Digital access controll and prevention parameters |
| 13 | Isolation | Isolation parameters |
| 14 | Forensic capabilities | Forensic parameters |
| 15 | Change management | Change management parameters |

structure. Each parameter group contains parameter sub-group (and the sub-groups can contain parameter sub-sub groups as well with an indefinite depth). With this tree-like organization setup we can scale up the parameter field and include additional evaluation parameters easily without to influence other parts of the parameter tree. As a total of 450 parameter records have been identified so far inside the parameter tree.

### 4.2.2  Security Evaluation Parameter Sub-groups

We have identified the following main evaluation parameter sub-groups within the security parameter category (shown in Table 2).

## 5  Cloud Assessment

The term assessment is by definition, to measure something or calculate a value for it. Although the process of producing an assessment may involve an audit by an independent professional, its purpose is to provide a measurement rather than to express an opinion about the fairness of statements or quality of performance. An assessment implies a less independent and more consultative approach. The outcome of the assessment should relate to the norms that were set for the task, product or event.

Commercial cloud providers have less and less interest to maintain transparent and comparable systems, because they think they not only need to attract and catch the users but they also need to do lock-in. For such reason they provide minimalistic information about what is happening behind the business scenes, not only to the users, but also to the authorities. Some of the user communities require secure and highly transparent cloud infrastructure and services due to legal regulations; others need to be compliant with special data management requirements. Business can easily overwrite cloud providers' all internal best practice rules. This can be dangerous and harmful both for cloud users and providers. Data management issues, data leakage incidents, breached SLAs can easily remain hidden without periodically investigating the internal driving mechanisms of the cloud provider's ecosystem. Regular cloud assessments based on systematically collected information can become a useful service both for existing cloud users and cloud service providers.

### 5.1  Cloud Assessment Procedure

For the assessment procedure we have defined an easy-to-follow guided process. The workflow of the developed assessment procedure is shown in Fig. 6. The IaaS Cloud assessment process is starting by the request of the Cloud Service Provider

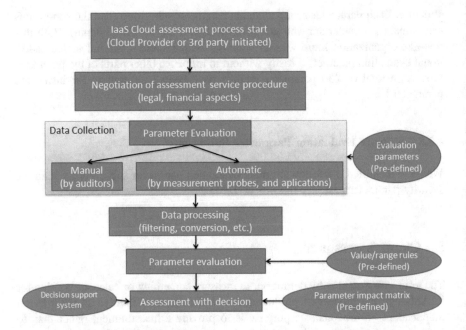

**Fig. 6** IaaS cloud assessment procedure

(owner of the infrastructure) or 3rd party (e.g. governmental) organization. The main elements of the assessment procedure are the following:

1. Negotiation phase: In this phase the preliminary assessment service require-ments (focus of the assessment, special assessment targets), financial, and legal aspects have to be identified.
2. Data collection phase: In this phase data is collected focusing on the existing cloud infrastructure, existing services, and existing procedures of the Cloud Service Provider. The evaluation parameter set is pre-defined. Data collection is realized by the auditors (manual data collection) and by measurement applica-tions, probes, tests (IT supported data collection). Manual data collection is maintained by the auditor person during the precise, on-site checking process The IT supported data collection is realized by Cloudscope. The internal architecture of the Data Collection framework enables easy integration of var-ious measurement tools.
3. Data processing phase: In this phase, each of the collected data sets are filtered, processed and converted to a single value, which is stored as the evaluation parameter result.
4. Parameter evaluation phase: In this phase each parameter of the whole evalu-ation parameter set is processed one-by-one by pre-defined evaluation rules.
5. Cloud assessment phase: In his phase the evaluated parameter set is multiplied by the parameter impact matrix, and according to the predefined decision rule the cloud provider receives the official assessment result.

# 6  Conclusions

In this paper, we discussed the major vulnerability sources in distributed computing infrastructures and we presented some of the recently used monitoring solutions and vulnerability frameworks for DCI's. We have designed and implemented a software solution called "Cloudscope", which is a modular security assessment framework and based on various existing security scanner solutions. We have incorporated a large number of external security solutions in a virtualized manner and build up a centralized security system monitoring user interface to show security assessment results for the administrators/users in a unified way.

The vulnerability testing module orchestrates remotely, all the loosely coupled scanning tools (e.g. PAKITI, GSSVA, AVAT, OpenVAS).

Each of these software solutions has been already used successfully by us to evaluate different type of DCI's (for example SEE-GRID-SCI's grid infrastructure/GLite based/, MTA SZTAKI's cloud infrastructure (Open Stack, Open Nebula based/, HP-SEE's supercomputing infrastructure/ARC based/. Using our framework, the combined vulnerability scanner tool set successfully revealed a large number of valid security problems and vulnerabilities with high impact on the examined infrastructures. With the forwarded detailed reports existing vulnerabilities have been eliminated by the local experts/administrators. The feedbacks received from the regional DCI administrators (from HP-SEE—supercomputing infrastructure, SEE-GRID-SCI—grid infrastructure, and various cloud service providers) proved, that it is a handy tool to make DCI's more secure. Furthermore, we have defined and built up an exhaustive tree-like graph of cloud evaluation parameter sets (our set contains more than 450 evaluation parameters). We showed the core parameter categories and as an example one of the sub-parameter sets (the security parameter group) to detail how the whole tree is organized. We have defined a generic core service assessment workflow and have identified its main tasks and additionally the relations between the tasks. We have used the defined parameter tree and the evaluation workflow in a cloud infrastructure related R&D project focusing on cloud providers. With the whole parameter tree we are able to realize IaaS type cloud service and infrastructure assessment. As an output we are able to use the defined processes and evaluation parameters to initiate a cloud assessment service both for cloud service providers and cloud customers.

# 7  Future Work

In future work we are planning to add to Cloudscope additional software components designed to evaluate additional aspects such as security compliance. We are also planning to further expand the security/vulnerability assessment toolset of the framework.

**Acknowledgment** The research leading to these results has received funding from the European Social Fund and the Hungarian TÁMOP-4.2.1.B-11/2/KMR-2011-0001 "Kritikus infrastruktúra védelmi kutatások" project. Authors would like to thank for the helpful technical support of the Laboratory of Parallel and Distributed Systems (LPDS) at MTA SZTAKI.

# References

1. http://www.eucalyptus.com. Accessed 10 Feb 2014
2. Bresnahan, J., LaBissoniere, D., Freeman, T., Keahey, K.: Cumulus: An Open Source Storage Cloud for Science, ScienceCloud 2011, San Jose, CA. June 2011
3. Sotomayor, B., Montero, R.S., Llorente, I.M., Foster, I.: Virtual infrastructure management in private and hybrid clouds. IEEE Internet Comput. **13**(5), 14–22 (2009)
4. http://www.openstack.org. Accessed 10 Feb 2014
5. Ács, S., Kozlovszky, M.: Advanced Vulnerability Assessment Tool for Distributed Systems; HP-SEE User Forum 2012, BoA. pp. 46. Belgrade, Serbia, 17–19 Oct 2012
6. Bogan, C.E., English, M.J.: Benchmarking for Best Practices: Winning Through Innovative Adaptation. McGraw-Hill, New York, NY
7. Martin, R.A.: Managing Vulnerabilities in Networked Systems. IEEE Computer Society COMPUTER Magazine, pp. 32–38 (2001). http://cve.mitre.org/
8. Mell, P., Scarfone, K., Romanosky, S.: A complete guide to the common vulnerability scoring system, version 2.0. Forum of Incident Response and Security Teams, June 2007
9. http://aws.amazon.com/cloudwatch/. Accessed 10 Feb 2014 February
10. Acs, S., Kozlovszky, M., Balaton, Z.: Automation of security analysis for service grid systems. In: Topping, B.H.V., Iványi, P. (eds.) Proceedings of the First International Conference on Parallel, Distributed and Grid Computing for Engineering, Civil-Comp Press, Stirlingshire, UK, Paper 25, 2009. doi:10.4203/ccp.90.25, ISSN 1759-3433
11. http://pakiti.sourceforge.net/. Feb 2013
12. The OpenVAS website, http://www.openvas.org. Accessed 10 Feb
13. http://www.tenable.com/products/nessus. Accessed 10 Feb 2014
14. http://www.rapid7.com/products/nexpose/. Accessed 10 Feb 2014
15. http://www.rapid7.com/company/news/press-releases/2012/usgcb-cyberscope.jsp. Accessed 10 Feb 2014
16. http://www.metasploit.com/. Accessed 10 Feb 2014
17. http://en.wikipedia.org/wiki/Microsoft_Baseline_Security_Analyzer. Accessed 10 Feb 2014
18. http://www.qualys.com/enterprises/security-compliance-cloud-platform/. Accessed 10 Feb 2014
19. https://www.netiq.com/products/sentinel/. Accessed 10 Feb 2014
20. Kozlovszky, M.; Trocsik, M.; Schubert, T.; Poserne, V.: IaaS type cloud infrastructure assessment and monitoring. In: 2013 36th International Convention on Information & Communication Technology Electronics & Microelectronics (MIPRO), pp. 249,252, 20–24 May 2013
21. Cloud Security Alliance—CSA: https://cloudsecurityalliance.org/. Accessed 10 Feb 2014
22. European Network and Information Security Agency—ENISA: http://www.enisa.europa.eu/. Accessed 10 Feb 2014
23. The Opengroup Jericho Forum: http://www.opengroup.org/getinvolved/forums/jericho. Accessed 10 Feb 2014
24. The Federal Risk and Authorization Management Program (FedRAMP): www.fedramp.gov. Accessed 10 Feb 2014
25. http://sas70.com/. Accessed 10 Feb 2014

26. http://ssae16.com/SSAE16_overview.html. Accessed 10 Feb 2014
27. http://www.aicpa.org/InterestAreas/FRC/AssuranceAdvisoryServices/Pages/SORHome.aspx. Accessed 10 Feb 2014
28. https://cloudsecurityalliance.org/download/security-guidance-for-critical-areas-of-focus-in-cloud-computing-v3/. Accessed 10 Feb 2014

# The Effect of Comminution as a Pretreatment Method Used in the Process of Anaerobe Fermentation of Lignocellulose Substrate on Biogas Yield

**Miklós Horváth**

**Abstract** Biogas technologies which are based on anaerobic fermentation, have to face several problems today. The general opinion of research institutes is that primary utilization of biogas plants would be a reasonable waste management and the energy produced should be seen as a secondary benefit. However, with targeted research and development a significant increase can be achieved in gas yield due to pretreatment technologies and process optimizations, i.e. biogas technologies have still too many reserves. In this paper, two pieces of mechanical equipment will be discussed, which are suitable for certain lignocellulose substrate pretreatment before anaerobic fermentation. As the effect of pretreatment the carbon/nitrogen ratio of the fermenter and the quantity of the biogas produced during the same time can be increased and also an improvement can be achieved in the context of problems arising from floating substrate.

**Keywords** Pretreatment · Comminution · Biogas · Lignocellulose

## 1 Introduction

### 1.1 Biogas as a Renewable Energy Source

Biogas is gaseous material, which is very similar to natural gas. It has a composition of 50–70 % methane, 28–48 % carbon dioxide and 1–2 % of other gases, mainly hydrogen sulfide and nitrogen [1]. Biogas generation is a process occurring independently in nature (e.g. Marsh gas, ruminant digestive system, etc.), however, in laboratories and industrial environment the anaerobic fermentation of organic

M. Horváth (✉)
Faculty of Mechanical and Safety Engineering, Óbuda University, Budapest, Hungary
e-mail: horvath.miklos@bgk.uni-obuda.hu

© Springer International Publishing Switzerland 2016
L. Nádai and J. Padányi (eds.), *Critical Infrastructure Protection Research*,
Topics in Intelligent Engineering and Informatics 12,
DOI 10.1007/978-3-319-28091-2_12

141

matter is carried out in the presence of microorganisms. According to practical experience, any organic material can be used as substrate, if approximately 230–400 l biogas can be produced from its all dry matter kilograms. The seed sludge, the substrate used and the technology may cause the difference [2]. In addition in the fermentation process, carried out by microorganisms, valuable manure is generated, which represents nearly twice as much specific nutrient content, than livestock manure [1]. Regarding its structure and composition fermentation residual can improve soil activity, which can be used especially in poor soil types, (e.g.: increasing humus).

Today, the biogas production on the market basis does not fall within the scope of low-cost energy production, still not negligible. The rising extraction costs of fossil fuels and dependence of many countries on fossil energy have increased the need for alternative energy sources, and so for biogas technologies [3]. Although the energy produced in this way is relatively expensive, and obtaining the necessary raw material also causes increasing difficulties, effective waste management, related by-products and the secondary energy produced, together moving towards sustainable development. The role in perception of biogas production can be improved mainly through research and development in the field of process optimization and pretreatment because the technologies have still significant reserves. In Western Europe, primarily Germany is the most important user of biogas technology. Nearly 8000 biogas plants are in operation in Germany today, mainly due to the strong political support of renewable energy sector, on the other hand, due to compensatory state aid for the higher cost of alternative energy sources. In Hungary, the biomass-based energy production represents only a small portion of the energetic use of biogas from fermentation. According to the survey made by the Subcommittee on Renewable Energies, Energy Commission of Hungarian Academy of Sciences (MTA) in 2003–2005, the theoretical energy potential of biomass of Hungary was 203.1–328 PJ/year [4], while according to a study made by the Ministry of Agriculture and Subcommittee on Renewable Energy Technologies of Hungarian Academy of Sciences, in 2005, the estimated total biogas potential was 25–48 PJ/year.

## 1.2 Types of Biogas Plants

Biogas plants can be grouped by

- The dry matter content of raw materials (wet, semi-dry)
- The method of operation (continuous, intermittent, combined)
- The duration of organic matter degradation (2 days–25 years)
- The construction, or arrangement (vertical, horizontal, tube)
- The temperature range used in the fermentation (psychrophilic, mesophilic, thermophilic)

**Fig. 1** Agricultural raw material processing biogas plant (Pilze-Nagy Ltd., Kecskemét). *Source* own photo

- The raw materials used
- Other aspects [2]

Three main plant types can be distinguished according to the raw materials used [5]. Experience has shown that this division can be used on the global level (e.g. the American Biogas Council [6]), and also in Hungary. There are overlaps, but agricultural raw material processing (Fig. 1), wastewater treatment plants (Fig. 2) and landfill gas recovery plants can be distinguished [7]. Compared with the first two the landfill gas recovery plant shows a significant difference, because technically there's no raw material input. This plant consists of a vertical pipe system drilled into the closed landfill and of their collecting points (Fig. 3); the gas is used on site, or within an area where it can economically be transported. Agricultural raw material and/or waste processing biogas plants consist of low-designed digesters (fermenters), where the substrate is continuously fed to the seeding sludge. There are many types of substrates from livestock manure to silage maize.

The tall designed fermenters located on wastewater treatment plants mainly serve to limit the amount of excess sludge; the surplus energy is good use to the operation of the plant and its environment [5].

## 1.3 The Role of Pretreatment in Biogas Technology

Biogas generation takes place in anaerobic environment as a result of the coordinated operation of a number of microorganisms. The process consists of three steps,

**Fig. 2** Wastewater treatment plant, high digester tower. *Source* own photo

**Fig. 3** Collecting point of a gas well pipe network on a landfill. *Source* own photo

which cannot be separated from each other in natural environment. Hydrolyzing organisms "cut" the polymers with their exoenzymes. The resulting smaller molecules are utilized as a nutrient by bacteria. The extracellular hydrolytic enzymes prepare the hydrolysis; degrade all available molecules, thus providing nutrients to the hydrolyzing bacteria and other micro-organisms. The hydrolyzing bacteria continue to metabolize nutrients, using the stored energy released during the breakdown of chemical bonds. The resulting essential organic acids (e.g. butyric acid, acetic acid, propionic acid) are released into the environment.

The second main step of the process is when the acetogenic bacteria convert the previously unused oligo-and monosaccharides, amino acids and fatty acids, and also, the essential organic acids into hydrogen and acetate.

In the third stage of the process, the methanogenic microorganisms, called methanogens produce biogas, the gas mixture containing mainly methane and carbon dioxide. Some methanogens reduce the carbon dioxide with hydrogen, while others are able to convert the acetate to methane and carbon dioxide.

Besides a very brief summary of the process it is appropriate to mention that it is by no means about independent systems, but also about close, interdependent microbiological relationship chain, in which not fully known processes play a role, such as "hydrogen transfer" mechanism [1].

It is the biogas yield that shows the efficiency and profitability of the anaerobic treatment. The activity of microbes responsible for the biogas yield can be improved by better understanding the microbial processes described above, and by improving the digestibility of wastes. The latter option includes pretreatment, which results in better access to the nutrients for bacteria [8].

Because of the biogas yield the accessibility is important in itself, but there are also other important aspects that may make the application of pretreatment necessary.

To build up the microbial cell proteins nitrogen is needed in the fermentation process. If the amount of nitrogen is not sufficient to process the large amount of carbon is inhibited. If there is too much nitrogen, ammonia accumulates, which inhibits the growth of bacteria and eliminates anaerobic conditions. It follows that appropriate C/N ratios (20–30) are needed for optimal conditions. To set this value stirring and proper pretreatment of various materials are required.

Table 1 shows the C/N ratio of various substrates. Some livestock manures (e.g. poultry litter), are rich in nitrogen that requires an increase of carbon content for optimal biogas yield. One of the most obvious solutions is the utilization of locally available wheat straw in the fermenters. However, without pretreatment, the straw floats up blocking the appropriate stirring and the straw forming a coherent layer on

**Table 1** C/N ratios of various substrates [2]

| Name of substrate | C/N ratio |
|---|---|
| Wood, sawdust | 200–500 |
| Straw | 50–150 |
| Communal waste | 30–40 |
| Hay, grass, green waste | 10–20 |
| Cattle, swine, poultry manure | 5–10 |
| Algae, microbiological residues | 5–10 |
| Slaughter-house waste | 2–4 |

the top of the fermenter blokes the path of methane as well. Comminution always means energy investment, but experience has shown it is not enough, as the simple length reduction—although increases surface—does not provide sufficient access to the cellulose chains for bacteria. In laboratory experiments it was proved that the use of a cavitation field further increases the surface of the previously comminuted wheat straw [9]. Direct destruction of the surface improves the efficiency of comminution, in the framework of complex pretreatment. Cavitation is the formation and violent collapse of bubbles in liquid medium, generating high temperature (about $10^4$ K) and pressure (about $10^3$ Bars) conditions. Due to the non-symmetrical collapse of the bubbles shock waves are formed in the liquid [10]. Since this phenomenon occurs in a high temperature and pressure, erosion/corrosion develops on the surface of different particles in the liquid material [11]. The generated shock waves result in a significant acceleration of particles as well [12, 13].

## 2 Experimental Apparatus, Materials and Methods

### 2.1 Experimental Apparatus

During pretreatment experiments the combined effect of two devices was studied. Figure 4 shows a dry crusher (Szanyi 600 straw milling equipment) as a first step after harvesting straw. The straw is manually fed into the crusher. The power to the operation of the device provided by an 11 kW electric motor, which drives the grinder—shredder unit and a flywheel, too. The straw is crushed by the rotating and stationary blades of the machine, and then with an approximately homogeneous fraction size it gets to a milling device, where between 6 × 4 hammers and a screen the pre-shredded material undergoes additional shredding, until the particles pass through the screen. The number of rotations: 1450 1/min, shredding capacity: 600–1500 kg/h.

The Hydrodynamic cavitation device (HCD—Fig. 5) is driven by a 7.5 kW electric motor. The equipment consists of an electric motor, which drives a solid disc (diameter 150 mm, width 150 mm, preferably made of stainless steel) directly through a shaft. On the surface of the disc there are holes of about 10 mm depth, in rows. The rotary disc is covered by a fixed cylinder (stator); the gap between them is 3 mm. The low-solid-content (2–3 %) straw-water suspension is fed from axial direction onto the surface of the cylinder, where getting into the holes there is a sudden pressure drop and cavitation develops. Then the suspension exits on the other side of the cylinder. The number of rotations used during the treatment with the HCD is up to 3000 rpm.

**Fig. 4** Szanyi 600 straw milling equipment. *Source* own photo

**Fig. 5** Hydrodynamic cavitation device. *Source* own photo

## 2.2 Materials and Methods

Comparative experiments were used to examine the effect of pretreatment. As a first step untreated (cut with scissors) straw was compared to the straw shredded with Szanyi 600, and then in a subsequent experiment, the latter was compared to the straw which was further treated with HCD. Air-dried wheat straw was used to the experiments. The comparative tests were carried out according to VDI 4630 recommendations [14], with batch experiments, in which "three-glass systems" were used in parallel measurements, in the presence of negative and positive controls, in mesophilic conditions. The three-glass systems were placed into a proprietary water bath and incubator cabinet, which are shown in Fig. 6. On the basis of VDI 4630 recommendations the seed sludge and the substrate are placed into the first bottle, from which the developing gas flows to the middle, water-filled bottle. As an effect of pressure difference the fluid moves to the third bottle. The volume of gas produced can be calculated from the amount of liquid flowing into the third bottle. The quality of the generated gas was established with DANI gas chromatograph, the analysis of the results was carried out with the help of Clar software.

The wheat straw shredded by Szanyi 600 got into a cyclone apparatus with the particle size of the used screen size, and after de-dusting, into a storage bag. Before the comparative experiments the straw was fractionated by a vibratory sieve shaker

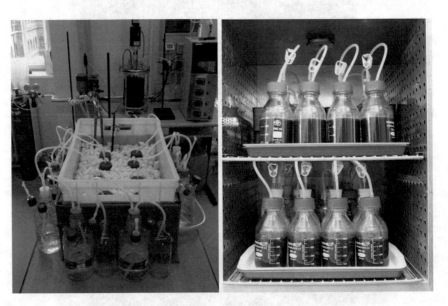

**Fig. 6** Batch reactor systems used during the experiments. *Source* own photo

(Retch AS 200). 0.125–1; 0.125–2 and 2–4 mm sieve dimensions were used, which refer to two dimensions of the straw. With an average length of 5, 10, and 20 mm, the third dimension is always longer.

Wheat straw shredded by Szanyi 600 and soaked in water for 10 min was used in the HCD device as raw material. The suspension was continuously circulated and treated at 2300 1/min for 6 min.

# 3 Results and Discussion

In the first and second experiment, on the first day the lower yield from the straw treated with Szanyi 600 presumably due to the minimum overfeeding which occurred at the starting of the system, and caused inhibited fermentation runs (minimum shock to the bacteria), but then a higher yield was observed in favor of the Szanyi 600-treated sample. At the end of the process the total gas yield is clearly higher with the Szanyi-600 pretreated sample (Figs. 7 and 8).

In the second experiment the smaller particle size range resulted in larger differences as the effect of pretreatment with Szanyi 600. In this case, the measured 19 % difference is due to the larger proportion of surface digestion at smaller particle size. It is worth mentioning that in both cases, the methane content showed higher growth, that is in the first case beside the 6.2 % biogas yield growth the methane yield growth was 13.3 %, while beside 19 % biogas yield growth it was 19.4 %.

The relatively large standard deviations result from the inhomogeneity of biogas systems. The standard deviation values, used in accordance with VDI 4630 for small scale experiences, are acceptable.

Due to the facilities and characteristics of the equipment, the conditions (mud, weighed dry matter content, grain size) were different in the third series of comparative experiments, but the 42.9 % growth of gas yield and the 48.3 % growth of methane yield shows a significant result in favor of the combined pretreatment [7].

# 4 Conclusions

There are two undeveloped areas of biogas producing technologies: the optimization of the technological process and the utilization of energetically profitable pretreatments. The experiments described above, demonstrate that even dry shredding can effectively increase the total gas yield, obtained at the same time and comparing the gas yield, in all cases, in a greater extent, the methane yield. With the

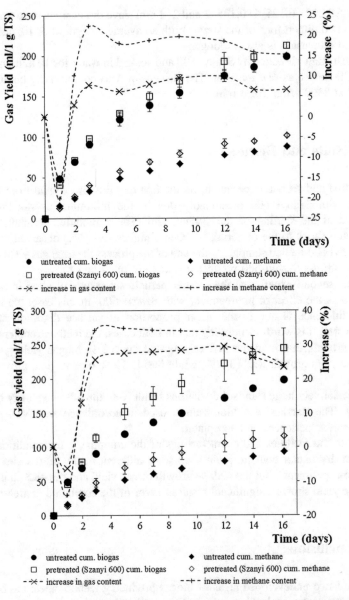

**Fig. 7** First and second series of comparative experiments—Biogas and methane yields in the function of fermentation time with untreated (cut by scissors) and pretreated with Szanyi 600 straw (particle size: 2–4 mm, and 0.125–2 mm); and increase of gas content comparing to untreated substrate. *Source* [7]

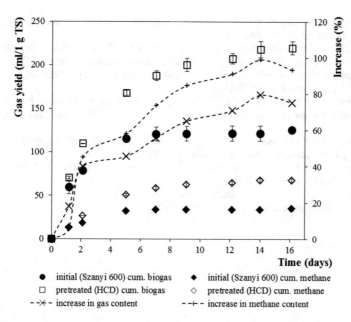

**Fig. 8** Third series of comparative experiments—Biogas and methane yields in the function of fermentation time with straw pretreated with Szanyi 600 (starting substrate) and with HCD (particle size: 0.125–1 mm); and increase of gas content comparing to starting substrate. *Source* [7]

combined pretreatment method, i.e. by further treatment of the dry-shredded straw using the destructive effects of cavitation, further significant growth can be achieved in both, the biogas and the methane yields.

**Acknowledgements** Special thanks to Márton Szigeti for the active co-operation in experiments, to Gabor Nyikos for making available Szanyi 600. Special thanks to Monika Bakosné Diószegi, Levente Csóka and Tibor Poos for the help in the development of equipment and experiments.

# References

1. Bai, A.: A Biogáz. Száz Magyar Falu Könyvesháza Kht, Budapest (2007)
2. Drégelyi Kiss, Á., Horváth, M., Bagi, Z.: Biogáz gyártás mérési eljárásai, Óbudai Egyetem, Budapest (2012)
3. Bakosné Diószegi, M.: A magyarországi energiabiztonság növelésének okai. In: *XXI. Nemzetközi Gépészeti Találkozó*, pp. 28–31. Erdélyi Magyar Műszaki Tudományos Társaság, Kolozsvár (2013)
4. Tóth, P., Bulla, M., Nagy, G.: Energetika, Digitális Tankönyvtár (www.tankonyvtar.hu) (2011)
5. Horváth, M., Horváth, S., Bakosné Diószegi, M., Poós, T.: "Breakthrough possibilities and limitations based on the experiences of the Hungarian biogas plants. In 5th International Symposium on Exploitation of Renewable, pp. 123–126. Subotica (2013)

6. American biogas council, [Online]. Available: http://www.americanbiogascouncil.org/biogas_ maps.asp. Accessed 21 April 2014
7. Horváth, M.: Increasing biogas yield by combined milling processes. In International Engineering Symposium at Bánki, pp. 205–212. Óbudai Egyetem. Budapest (2013)
8. Bakosné Diószegi, M., Horváth, M., Szigeti, M.: Biogáz alapanyag előkezelési technológiák. Óbudai Egyetem, Budapest (2013)
9. Iskalieva, A., Mbouyem Yimmou, B., Gogate, P.R., Horváth, M., Horváth, P.G., Csóka, L.: Cavitation assisted delignification of wheat straw: a review. Ultrason. Sonochem. **19**, 984–993 (2012)
10. Leighton, T.G.: The Acoustic Bubble, pp. 531–551. Academic, London (1994)
11. Preece, C.M., Hansson, I.L.: Adv. Mech. Phys. Surf., **1**:199 (1981)
12. Suslick, K.S., Doktycz, S.J.: The effects of ultrasound on solids. In: Mason, T.J. (eds) Advances in Sonochemistry, pp. 197–230 (1990)
13. Doktycz, S.J., Suslick, K.S.; Science, p. 1067 (1990)
14. V. 4630, Fermentation of organic materials (2006)

# Experimental Investigation of Stress Distribution in a Tensile Test Specimen, Using a Novel Gripping System for Tensile Testing

**András Mucsi**

**Abstract** This article deals with the experimental verification of the stress distribution inside a tensile test specimen using a recently developed gripping system. The upper yield strength of materials such as low carbon steels could be extremely sensitive to the elastic stress distribution inside the test piece. The evolution of stress distribution inside the sample during elastic loading is in strong correlation with the initial alignment and gripping methodology of the test piece, so the gripping system influences strongly the value of upper yield strength. Therefore, the application of a suitable gripping system is essential if the exact value of the upper yield strength is necessary to know. In the current work a special test piece equipped by strain gauges has been used to measure the stress distribution during elastic loading of the tensile test piece. The results show that the new gripping system could provide almost pure uniaxial conditions; moreover the uniaxial or even non-uniaxial loading can be adjusted and repeated with small deviation.

**Keywords** Tensile test · Gripping system · Upper yield strength · Strain gauge

## 1 Introduction

Tensile testing of materials plays a general role in industrial practice. Numerous studies have been published in the topic of tensile testing, gripping systems and interpretation of test results. A comprehensive study about tensile testing can be found in the work of Davis [1].

The upper yield strength of materials is one of the most important properties, because it is considered as the stress level where the large-scale plastic deformation

A. Mucsi (✉)
Donát Bánki Faculty of Mechanical and Safety Engineering,
Óbuda University, Budapest, Hungary
e-mail: mucsi.andras@bgk.uni-obuda.hu

© Springer International Publishing Switzerland 2016
L. Nádai and J. Padányi (eds.), *Critical Infrastructure Protection Research*,
Topics in Intelligent Engineering and Informatics 12,
DOI 10.1007/978-3-319-28091-2_13

begins. According to ISO-6892 the upper yield strength is defined as the stress where the force first time decreases during a tensile test (it reveals to the beginning of plastic flow). The value of it strongly depends upon the circumstances of tensile test. It was shown that the gripping system and the initial alignment of the test piece could strongly influence its value [1–6].

The main problem of measuring the upper yield strength of materials is caused by the eccentricity between resultant loading force and the axis of symmetry of test piece. If a small eccentricity is arising between the resultant loading force and the axis of symmetry of test piece during the initial (elastic) stage of tensile test, the summarized stress will be larger in a distinct part of the specimen than it is expected from the loading force/cross sectional area ratio. This phenomenon results in an apparent decrease in upper yield strength [1, 2].

Beside the significance of the axiality of the loading of a test piece during elastic loading of a tensile test specimen at ambient temperature, the gripping system also plays an important role during high temperature creep tests [7, 8]. The specimen used for a creep test is loaded only elastically as well as the tensile test piece at the beginning of the test. This is the reason that the gripping system and the initial alignment of the test piece play a decisive role by both testing procedure. As soon as large plastic deformation occurs, the test piece will align itself to the uniaxial conditions and no effect of the initial misalignment can be obtained in the remained part of the test [2].

The stress distribution inside a tensile test piece gives reliable information about the applicability of the gripping system for measuring the upper yield strength. Numerous methods have been developed to estimate the stress distribution in engineering parts. Beside the widely used strain-gauge technique [7–14], some other methods are generally accepted such as neutron diffraction [10], interferometry [15] and finite element analysis [12].

In the topic of "stress distribution in the tensile test specimen" only a few experimental studies have been published [3, 7–9]. The effect of wedge and screw grips on the stress distribution in a tensile test piece is clearly demonstrated in Gray and McCombe [3] earlier work. They used round specimens and strain gauges mounted onto the test piece in order to investigate the bending stress arising in the sample due to tensioning. They found, that the bending stress caused by screw grips and wedge can reach to 80 % of the average stress, moreover the diagrams represent that at the center of the specimen the stress concentration is smaller than near the ends. Because of it the plastic flow and propagation of Lüders bands sometimes starts near the grips [1–3, 6, 16].

In order to avoid the errors caused by the improper alignment of a test piece, a new gripping system was developed [2]. The new gripping system provides practically uniaxial loading if the eccentricity (defined below) is held nearly zero. Using this new gripping system, the upper yield strength and the "extrapolated upper yield strength" can be estimated with a high accuracy. The assembly of the new gripping system and the definition of eccentricity is shown in Fig. 1. For detailed description of the new gripping system see the earlier study of the author [2].

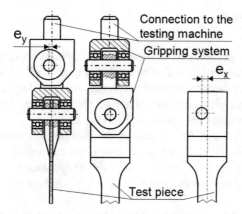

**Fig. 1** The gripping system tested in this study and the definitions of eccentricities [2]

The aim of the current study is to demonstrate the suitability of the gripping system to true uniaxial loading. The axial and bending stresses during elastic loading of the test piece were measured. The stress distribution measured at different X-direction eccentricities ($e_x$) but at constant $e_y = 0.02$ mm eccentricity in Y direction. The eccentricities shown in Fig. 2 were measured using a Mitutoyo PJ-H3000F profile projector.

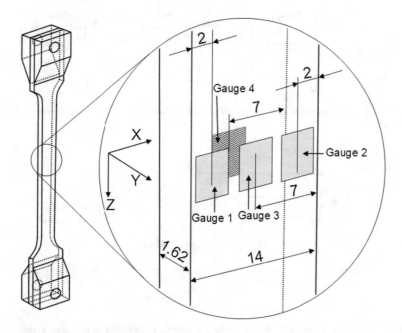

**Fig. 2** Test piece set up for estimating the stress distribution

## 2  Experimental

The aims of the experiments were to evaluate the average bending stress/axial stress ratio at different eccentricities. For measuring the stress distribution inside the test piece, a rectangular low carbon steel specimen was used. Strain gauges were fixed using high strength glue onto the center of the specimen. The positions of the gauges on the specimen are represented in Fig. 2.

Gauge 1 and 2 serve for measuring the deformation caused by bending in direction X, whilst Gauge 3 and 4 are designated to measure the deformation in direction in Y. The electrical connection of the gauges and the electrical circuit for signal conditioning are shown in Fig. 3. $U_t = 5$ V has been used for supplying the bridge.

In order to avoid the errors caused by a change in temperature, the compensation resistances $R_{1...4}$ were the same type of strain gauges as Gauge 1–4. They are located nearby Gauge 1–4 but they are not fixed onto the specimen. The quantitative evaluation of the bending stress from voltage data is based on the calibration curve. For taking up the calibration curve, the specimen is leaned at the fortification's hole in horizontal position and loaded at the center of it with weights. The graphic of this procedure is represented in Fig. 4.

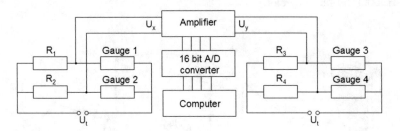

**Fig. 3** Electrical connection of strain gauges used for evaluating the stress distribution inside the test piece

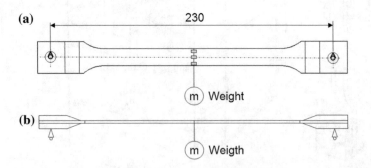

**Fig. 4** The sketch of the calibration procedure in direction X (**a**) and in direction Y (**b**)

The bending stress ($\sigma_x$, $\sigma_y$) arising due to the elastic loading of the specimen with standards according to Fig. 4 can be calculated in direction X as:

$$\sigma_x = \frac{M_b}{I_x} \cdot p \tag{1}$$

and in direction Y as:

$$\sigma_y = \frac{M_b}{I_y} \cdot p \tag{2}$$

where $M_b$ is the bending moment, F is the loading force ($F = m \cdot g$), m is the mass of the standard, g is the gravitational acceleration, $g = 9.81$ ms$^{-2}$); p is the distance of gauges from neutral line, $I_x$ and $I_y$ is the secondary moment of inertia in direction X and Y, respectively. The calibration curves are represented in Fig. 5.

The calibration constants were found to be 47.013 MPa/V for bending in direction X and 40.316 MPa/V for bending in direction Y. Using these calibration constants the bending stress during axial loading can be calculated as:

$$\sigma_x = c_X \cdot U_x = 47.013 \cdot U_X \tag{3}$$

in direction X, whilst in direction Y the following expression is valid:

$$\sigma_y = c_Y \cdot U_y = 40.316 \cdot U_Y \tag{4}$$

where $c_x$ and $c_y$ are the calibration constants in direction X and Y; $U_x$ and $U_y$ are the measured voltages generated by the strain gages in direction X and Y, respectively. The accurate measurement of axial stress is also necessary. The measurement of the axial loading force provides the evaluation of the average axial stress inside the test piece:

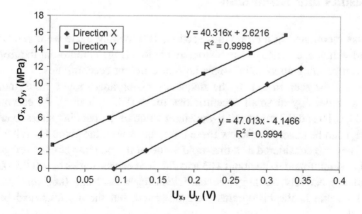

**Fig. 5** Calibration curves for evaluating the bending stresses

$$\sigma_a = \frac{F}{A_0} = \frac{c_F \cdot U_F}{A_0} \tag{5}$$

where $c_F$ is the calibration constant of the loading cell ($c_F = 10480$ N/V), $A_0$ is the cross-sectional area of the specimen, $U_F$ is the voltage generated by the loading cell due to loading.

The bending stress/axial stress ratio in direction X and Y were calculated as:

$$H_X = \frac{\sigma_x}{\sigma_a} \cdot 100\% \tag{6}$$

$$H_Y = \frac{\sigma_y}{\sigma_a} \cdot 100\% \tag{7}$$

Using the special test piece introduced above the probability function of the bending stress/axial stress ratio has been measured at five different eccentricities in direction X: $e_x = 0.03, 0.16, 0.3, 0.42$ and 0.75 mm. In direction Y only one eccentricity was used: $e_y = 0.02$ mm. At each individual eccentricity 100 axial loading cycles were performed using a TTM-100 electromechanical tensile testing machine with 5–10 MPa/s loading rate. During each loading cycle the bending stress (caused by the non-axial loading) in direction X and Y and the axial stress, such as their ratio have been calculated. The maximal value of the applied axial stress during the consecutive loadings had an approximately uniform distribution between 70 and 150 MPa. The lower yield strength of the low carbon steel used for the test piece was previously measured on other specimens made of the same material as the test piece according to ISO-6892 and a value of $R_{eL} = 238 \pm 7$ MPa was obtained. It means that the special strain gauge mounted test piece was loaded only elastically.

## 3   Results and Discussion

As it was mentioned in the previous section, 100 loading cycle were performed at each individual eccentricity. The histograms below (Fig. 7) represent the frequency of a given bending stress/axial stress ratio at different eccentricities.

As it can be seen in Fig. 6, the frequency histograms show a nonsymmetrical behavior, especially at small eccentricities ($e_x = 0.03$–0.3 mm). At eccentricities $e_x = 0.03, 0.16$ and 0.3 mm a quasi-threshold value of the bending stress/axial stress ratio ($H_x$) can be observed. Below these the frequency of the given $H_x$ value is very small. They are considered as a threshold value of the bending stress/average stress ratio $H_x$, which found to be about 1, 3 and 7.4 % at eccentricities $e_x = 0.03, 0.16$ and 0.3 mm, respectively. At larger eccentricities ($e_x = 0.42$ and 0.66 mm) no sharp threshold value in the histograms were observed, but the asymmetrical behavior remained.

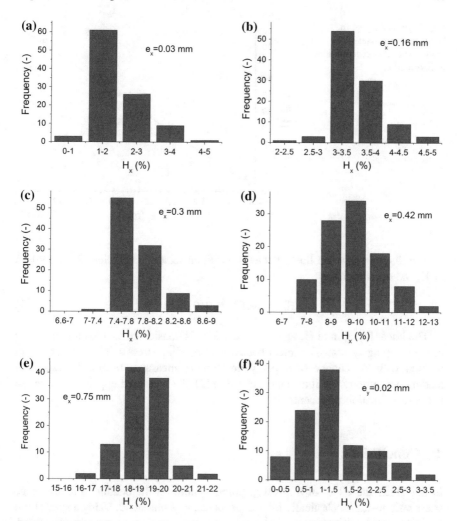

**Fig. 6** Frequency of the bending stress/axial stress ratio at different eccentricities in X-direction (a–e) and in Y-direction (f)

In direction Y only one eccentricity was studied. At $e_y = 0.02$ mm an average value of bending stress/average stress ratio $H_{y,avg} = 1.34$ % was obtained. The results exhibit nonsymmetrical distribution, but no sharp limit in the $H_y$ values was observed.

The measurements show a narrow deviation for the $H_x$ and $H_y$ values. The average and the deviation of $H_x$ values are shown against eccentricity $e_x$ in Fig. 7.

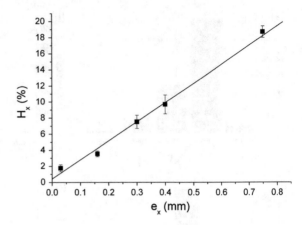

**Fig. 7** Frequency of the bending stress/axial stress ratio at different eccentricities in X-direction

The diagram shows a linear behavior (with correlation coefficient $R^2 = 0.9939$) of $H_x$ vs. eccentricity $e_x$:

$$H_x = 24.12 \cdot e_x + 0.3272 \tag{8}$$

The linear function of $H_x$ against eccentricity is considered as a loading property of the gripping system. The linear behavior of bending stress axial stress ratio is in relation with the change of upper yield strength measured by tensile tests with loading eccentricity. As it is demonstrated in [2], the measured upper yield strength is a linear function of eccentricity.

## 4  Conclusions

In this study the effect of a novel gripping system on the bending stress/average stress ratio inside an elastically loaded test piece, was measured. Using a special test piece equipped with strain gauges the bending stress/axial stress ratio was observed. On the basis of the experiments the following results were obtained:

1. In contrast to the wedge and screw grips, the novel gripping system is suitable for uniaxial tensioning of tensile test specimens with smaller than 2 % bending stress/axial stress ratio.
2. Using the novel gripping system a given bending stress/axial stress ratio is adjustable and repeatable within 1–2.5 % accuracy.
3. The gripping system tested in this study provides a linear relationship between the bending stress/average stress ratio ($H_x$) and eccentricity ($e_x$):

$$H_x = 24.12 \cdot e_x + 0.3272$$

4. With regard to the results published in earlier work[2] and the measurements represented in this study, we can conclude that the novel gripping system is suitable for accurate measurement of upper yield strength and can also be applied successfully for creep tests.

**Acknowledgements** The project was supported through the assistance of the European Union, with the co-financing of the European Social Fund (TÁMOP-4.2.1.B-11/2/KMR-2011-0001). The author is also indebted to Gedeon Richter Talentum Foundation for its financial support.

# References

1. Davis, J.R.: Tensile Testing, 2nd edn, p. 5. ASM International, USA (2004)
2. Mucsi, A.: Effect of gripping system on the measured upper yield strength estimated by tensile tests. Measurement **46**, 1663–1670 (2013)
3. Gray, T.G.F., McCombe, A.: Influence of specimen dimension and grip in tensile testing steel to EN 10 002. J. Iron Steel Making **19**, 405–409 (1992)
4. Sun, H.-B., Kaneda, Y., Ohmori, M. et al.: Effect of stress concentration on upper yield point in mild steel. Mater. Trans. **47**:96–100 (2006)
5. Docherty, J.G., Thorne, F.W.: The phenomenon of tensile yield in mild steel and iron. Engineering **132**, 295–297 (1931)
6. Hutchinson, M.M.: High upper yield point in mild steel. J. Iron Steel Inst. **186**, 431–432 (1957)
7. Bressers, J.: A code of practice for the measurement of misalignment induced bending in uniaxially loaded tension-compression tests pieces. Report, Joint Research Centre Institute for Energy. Petten, The Netherlands (1995)
8. Grant, C.: Axiality of loading in the tensile test. J. Strain Anal. **7**, 261–265 (1972)
9. Webb, I.N.: A system for the axial loading of creep specimens. Report, Ministry of Defence Aeronautical Research, London (1977)
10. Coules, H.E., Cozzolino, L.D., Colegrove, P., et al.: Residual strain measurement for arc welding and localised high-pressure rolling using resistance strain gauges and neutron diffraction. J. Strain Anal. **47**, 576–586 (2012)
11. Faghidian, S.A., Goudar, D., Farrahi, G.H., et al.: Measurement, analysis and reconstruction of residual stresses. J. Strain Anal. **47**, 254–264 (2012)
12. Mollicone, P., Gray, T.G.F., Duncan, C.: Experimental investigation and finite element analysis of welding induced residual stresses. J Strain Anal. **47**, 140–152 (2012)
13. Kumar, A., Chaturvedi, S.K., Chaturvedi, V., et al.: Design studies and optimization of position of strain gauge. Int. J. Sci. Eng. Res. **3**, 1–4 (2012)
14. Ivetic, G., Lanciotti, A., Polese, C.: Electric strain gauge measurement of residual stress in welded panels. J. Strain Anal. **44**, 117–126 (2009)
15. Tjhung, T., Li, K.: Measurement of in-plane residual stresses varying with depth by the interferometric strain/slope rosette and incremental hole-drilling. J. Eng. Mater. Technol. **125**, 153–162 (2003)
16. ISO 6892-1:2009 Metallic materials—tensile testing at ambient temperature

# Thermal Image Processing Approaches for Security Monitoring Applications

András Rövid, Zoltán Vámossy and Szabolcs Sergyán

**Abstract** A time series collected by thermal cameras yields many application possibilities for security monitoring applications. The character of features involved in thermal images is different compared to the images acquired by conventional capturing devices; many methods applicable for conventional image processing cannot directly be applied for thermal images. New methods are introduced in HOSVD based representation of thermal images, and in application of resolution enhancement and data compression. The retrieval process of thermal image data-bases and useful feature descriptors were analyzed, as well as the special semi-automatic fusion techniques of thermal image sequences.

**Keywords** Thermal images · HOSVD-based processing · Image representation · Image retrieval · Thermal image fusion

## 1 Introduction

Nowadays, security is a crucial issue, many critical structures are monitored by different types of devices. The collected time series can be used to detect different types of events, patterns, etc., automatically. Depending on their type, numerous methods have been developed to detect and recognize them efficiently. In the case of stations and airports, for instance, the detection of sick people, recognition of

A. Rövid (✉) · Z. Vámossy · S. Sergyán
Óbuda University, Budapest, Hungary
e-mail: rovid.andras@nik.uni-obuda.hu

Z. Vámossy
e-mail: vamossy.zoltan@nik.uni-obuda.hu

S. Sergyán
e-mail: sergyan.szabolcs@nik.uni-obuda.hu

© Springer International Publishing Switzerland 2016
L. Nádai and J. Padányi (eds.), *Critical Infrastructure Protection Research*,
Topics in Intelligent Engineering and Informatics 12,
DOI 10.1007/978-3-319-28091-2_14

163

irregular types of luggage, identifying abnormal behaviors and many other events or patterns play a significant role to guarantee the highest level of security as possible.

Depending on the type of sensors, various events can be detected. Let us focus on a time series collected by thermal cameras. Sick people (having fever) for example can be detected by thermal cameras easily, however, to detect more specific patterns in the thermal image relating to some more specific type of diseases is a much more difficult task.

Since the nature of the features involved in thermal imaging is different compared to the images acquired by conventional capture devices, many methods applicable for conventional image processing cannot directly be applied for thermal images. Thus, new methods are highly desirable.

In this paper let us focus on thermal image processing methods applicable for security monitoring related applications. First of all, let us emphasize the correspondence matching, feature extraction and thermal image representation related problems which may form the basis of many security monitoring related applications.

This paper is organized as follows: In Sect. 2 the HOSVD based representation of thermal images and images sequences is described including the applications such as resolution enhancement and data compression, Sect. 3 deals with retrieval processes of thermal image databases, describing the useful descriptors, comparison measures, and some special behavior of thermal images is mentioned. Section 4 introduces a new semiautomatic thermal image registration system together with the possible applications and finally the conclusions are reported.

## 2 Tensor Product Representation of Multidimensional Data

As the representation of data plays significant role during their processing, let us first give a brief description of the approach applied to represent thermal images as well as sequences in this section. Let us consider multidimensional data obtained through the discretization of a multivariate smooth function. In this case, the data is for the function values at the nodes of a hyper rectangular grid, over which, the multivariate function is discrete. In the case of images, these data represent the intensities of pixels or in the case of image sequences, represent the intensity of pixels in the frames shifted in time, by some $\Delta T$ value. Let us first take a simple example to make clearer the importance of the form of data representation. The image compression or the extraction of specific features in the image can be performed more efficiently in the frequency domain (where the data are represented by frequency components involved in the input signal) than in the spatial one.

Let us first express a multivariate smoothing function with the help of specific univariate smoothing functions forming an orthonormal basis, i.e.

$$f(x_1, x_2, \ldots, x_N) = \sum_{k_1=1}^{I_1} \cdots \sum_{k_N=1}^{I_N} \alpha_{k_1,\ldots,k_n} w_{1,k_1}(x_1) \cdot \ldots \cdot w_{N,k_N}(x_N), \qquad (1)$$

where the system of orthonormal functions $w_{n,k_n}(x_n)$ can be chosen in a classical way by orthonormal polynomials or trigonometric functions in separate variables. The numbers of functions $I_n$ in (1) are significant, since they stand for the components we are working on. Besides the number of functions their type or characteristics is also significant, these two factors are tightly connected, which means that in case of using simple components (like for instance trigonometric ones) to express a multivariate function we need more from them to achieve the required approximation accuracy than in case of using more specific components. Let us focus on the latter case, i.e. when the components are specific to the input multivariate function. We can determine such components trough higher order singular value decomposition (HOSVD) of our multidimensional data denoted by tensor $\mathcal{A} \in \mathbb{R}^{I_1 \times \cdots \times I_N}$. It can be expressed in tensor product (TP) form, as follows [1]:

$$\mathcal{A} = \mathcal{D}_A \times_1 \mathbf{U}_1 \times_2 \mathbf{U}_2 \ldots \times_N \mathbf{U}_N, \qquad (2)$$

where tensor $\mathcal{D} \in \mathbb{R}^{I_1 \times \cdots \times I_N}$ stand for the core tensor having the same number of dimensions as $\mathcal{A}$ and orthonormal square matrices $\mathbf{U}_n$ contain the function values of univariate components (as shown in [2, 3]) $w_{n,k_n}(x_n)$ at discretization points (nodes) corresponding to the $n$th dimension. Based on the above considerations, a color image represented by tensor $\mathcal{I}$ can be expressed in TP form as follows:

$$\mathcal{I} = \mathcal{D}_I \times_1 \mathbf{U}_1 \times_2 \mathbf{U}_2 \times_3 \mathbf{U}_3, \qquad (3)$$

where the 1st a 2nd dimension corresponds to $x$ and $y$ coordinates while the 3rd dimension stands for the color components (RGB, HSV, etc.). We can also merge different types of color components to form one tensor, as for instance, RGBHSV. In this case the size of our tensor along its 3rd dimension will be 6. Thus during the processing we can consider various properties of the image at once. The decomposition of a 3 dimensional tensor can be followed in Fig. 1.

Similarly, for a color image sequence we can write:

$$\mathcal{I} = \mathcal{D}_I \times_1 \mathbf{U}_1 \times_2 \mathbf{U}_2 \times_3 \mathbf{U}_3 \times_4 \mathbf{U}_4, \qquad (4)$$

where the 4th dimension corresponds to time axes. By manipulating the components involved in matrices $\mathbf{U}_n$ promising results can be achieved in case of compression, resolution enhancement, feature extraction, etc.

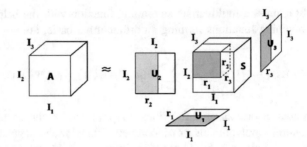

**Fig. 1** Illustration of the higher order singular value decomposition for a 3-dimensional array. Here **S** is the core tensor, the $\mathbf{U}_l$-s are the $l$-mode singular matrices

## 2.1 Resolution Enhancement of Thermal Images and Sequences

Based on the above considerations let us show how the resolution of an image or image sequence along whatever dimension can be enhanced by using the properties of HOSVD and tensor product form. Let $\mathbf{w}_n(x_n)$ denote a vector valued function defined as follows:

$$\mathbf{w}_n(x_n) = \left[ w_{n,1}(x_n), w_{n,2}(x_n), \ldots, w_{n,I_n}(x_n) \right]. \tag{5}$$

If it is discretized over the grid points corresponding to the $n$th dimension (see in previous section) its function values will stand for the rows of $\mathbf{U}_n$. Since the values of $w_{n,k_n}(x_n)$ are known at discretization points, the function value at any point falling into the discretization interval can be estimated by simple interpolation techniques. Thus the resolution enhancement can be performed in each dimension separately as follows:

$$\mathcal{I}_e = \mathcal{D}_I \times_1 \mathbf{U}_{1e} \times_2 \mathbf{U}_{2e} \times_3 \mathbf{U}_{3e}, \tag{6}$$

where the index $e$ denotes the enhanced form of the original tensors and matrices. Depending on the required resolution the size of matrices $\mathbf{U}_{ie}$ will vary. Similarly, in case of image sequences we can write the following:

$$\mathcal{I}_e = \mathcal{D}_I \times_1 \mathbf{U}_{1e} \times_2 \mathbf{U}_{2e} \times_3 \mathbf{U}_{3e} \times_4 \mathbf{U}_{4e}, \tag{7}$$

From (7) the enhanced frame corresponding to time $t$ can be expressed in TP form as follows:

$$\mathcal{I}_e = \mathcal{D}_I \times_1 \mathbf{U}_{1e} \times_2 \mathbf{U}_{2e} \times_3 \mathbf{U}_{3e} \times_4 \mathbf{w}_4(t), \tag{8}$$

The below example shows the efficiency of the approach for thermal image resolution enhancement (see Fig. 2).

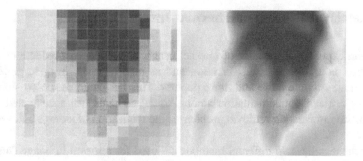

**Fig. 2** The original (*left*) and the enhanced (*right*) images [1]

## 2.2 Compression of Thermal Images and Sequences

It was shown in the previous section how the resolution can be improved in the case of images, as well as, images sequences. Let us now focus on the compression related possibilities of the HOSVD based representation. It is clear from (5) that the number of elements in the vector corresponds to the number of components in the $n$th dimension. Besides this, it is important to mention that these elements are ordered (from left to right) according to their importance. Thus, if some from the right most columns from matrices $\mathbf{U}_n$ are dismissed the tensor product will yield an approximation to the original image or image sequence. Since we are considering specific HOSVD based orthonormal components much more number of components can be dismissed than in case of using trigonometric or other basis functions. Consequently, by dismissing a given number of components from each dimension, efficient compression can be achieved. If the task is to compress the image sequence along the time axes (reducing the number of frames), can be achieved as follows:

$$\tilde{\mathcal{I}} = \mathcal{D}_R \times_1 \mathbf{U}_1 \times_2 \mathbf{U}_2 \times_3 \mathbf{U}_3 \times_4 \tilde{\mathbf{U}}_4, \tag{9}$$

where $\mathcal{D}_R \in \mathbb{R}^{I_1 \times I_2 \times I_3 \times r_4}$ denotes the reduced sized core tensor. Consequently the vector valued function in (5) will have a reduced number of elements, i.e.

$$\tilde{\mathbf{w}}_4(x_4) = \big[ w_{4,1}(x_4), w_{4,2}(x_4), \ldots, w_{4,r_4}(x_4) \big], \tag{10}$$

where $r_4 < I_4$. Similarly in case of applying the reduction in each dimension, the image sequence can be expressed as:

$$\tilde{\mathcal{I}} = \mathcal{D}_R \times_1 \tilde{\mathbf{U}}_1 \times_2 \tilde{\mathbf{U}}_2 \times_3 \tilde{\mathbf{U}}_3 \times_4 \tilde{\mathbf{U}}_4. \tag{11}$$

In this case $\mathcal{D}_R \in \mathbb{R}^{r_1 \times r_2 \times r_3 \times r_4}$ and $r_i < I_i$, $i = 1, 2, 3, 4$.

## 3   Retrieval Process of Thermal Image Databases

With the exponential increase of internet usage, as well as cost reductions in dig-
italization and storage assets, creation and storage of texts, images, graphs and
voices have become more and more popular. This has enlarged the possibility for
effective retrieval implementation among stored contents. An accentuated part of
the general information retrieval problem is the realization of storage and search of
images.

Fundamentally, two different methods are used for retrieval in image databases:
the text and content based approaches. The improvement of textually indexed
systems began on the years of 1970. In these systems textual descriptors are
manually assigned to images, and these descriptors ground for searching in the
database. This technique has two disadvantages. On the one hand, the implementation of textual indexing requires pregnant human resources. On the other hand, the
accuracy of textual indices depends on the subjective human perception. In the
interest of elimination of disadvantages of text based retrieval, in the 1980s began
the expeditious evolution of content based image retrieval systems. This evolution
is going nowadays as well.

In content based image retrieval systems the images are indexed by their own
visual content. The most frequently used features are color, texture and shape. In the
case of comparison of thermal images the color features cannot be used, since
thermal images can be considered as special gray-scale images. In this case the
comparison of images is very similar to the comparison of gray-scale images, but
the special properties of thermal images must be taken into account [4].

### 3.1   Intensity Descriptors

Due to description of gray-scale image intensity the following statistical moments
can be used. First order (mean), second order (variance), and third order (skewness).
These moments may be used as simple and effective characterization of intensity
distribution. The intensity histogram of a gray-scale image is defined as the
occurrence frequencies of each intensity value divided by the number of pixels in
the image.

Generally due to the comparison of intensity histograms, all possible intensity
values are not taken into account. The set of possible intensity values is divided into
subintervals (bins), and the histogram is expounded as the relative frequency of
pixels falling into these subintervals (bins).

It is well known that a small number of moments can characterize an image
fairly well, it is equally known that moments can be used to reconstruct the original
image. In order to achieve invariance to common factors and operations such as
scale, translation, and rotation, a set of moment invariants can be defined that we
use for the purpose of image retrieval of thermal images.

## 3.2 Edge Detection in Thermal Images

The goal is to find an edge matrix of a gray-scale image that fits best to the given image. Best fitting is, unfortunately, somewhat subjective. An often used method to find an edge matrix of a gray-scale image is to compute the gradient vector to each pixel. Then we say that a pixel is an edge pixel if the norm of its gradient exceeds a given threshold. The difference in edge detection algorithms lies in the interpretation of the gradient and in the selection of the threshold.

In the first step the noise reduction is our main goal. We have to select a noise reduction algorithm which does not significantly blur the edges (significant changes) from the image. As pre-processing the Symmetric Nearest Neighbor filter or the Kuwahara filter were applied.

The following steps of the algorithm [5] generate a metric to select the most appropriate threshold for edge detection. First, for each threshold value an edge matrix is produced. Those edge matrices where the number of edge pixels is less than the average of edge pixels in all matrices are dropped out. To enhance the remaining edge pixels, a morphological dilation is performed. At the finish of the algorithm, the threshold and the associated edge matrix are selected which are the most similar to all of the others.

## 4 Semi-automatic Thermal Image Registration System

The main goal of our system is to enable the user to examine a larger environment. However, the images are taken with a thermal camera, and not with a conventional camera, since it is appropriate for many applications to examine the objects based on their temperature in the workspace; typical applications are security systems, automatic detection of military targets, in case of poor visibility controlling outdoor space, or when thermal characteristics are relevant, such as the testing of thermal losses of a building.

The resolution of thermal cameras' is generally significantly lower than conventional cameras'. Therefore, it is rational to take a series of pictures of the environment and then to select two nearly identical images to find the best covered geometric transformation (homography transformation) that allows us to insert them into one picture. Matching the two pictures is automatic, but the selection of which of the two pictures to be chosen is up to the user. After the completion of the sequence alignment, we are able to study a larger working space thanks to the resulting mosaic image. During the operation, we assume that there are only spatial differences between the images, the recorded scene changes only minimally dynamically. Our earlier system [6] presented the principle of the usual steps of image registration [7]. However, intensities vary differently in thermal images than in the gray-scaled traditional camera images [8], thus it is essential at each step to provide the user with more possibility to execute the given task efficiently and to find the necessary partial methods. To make the system more reliable, users need to select from a variety of different techniques and to parameterize these methods.

## 4.1 The Principle of the Registration Process

The system solves a certain type of registration task when a series of pictures are taken by a thermal camera. Space part is inserted into one another on the basis of overlap and the feature sections are located there. For this reason, we try to find the special features that appear in the two images and are at the same part of the mapping object. These features appear in different places, though. The purpose of the registration is to determine the homography transformation, which maps the characteristic points (or feature points or corner points) of one image into the other image. First, we modify the image with the help of the transformation, and then we fuse this partial result to the original sample image. After the two joined images, selected by the user are loaded, four of the following steps are needed in order to prepare the fusion [7]:

1. Feature detection
2. Feature matching
3. Transformation model estimation
4. Image resampling and transformation

As the system has addressed registration for indoor and outdoor environments and working under different conditions, in each above mentioned steps several methods were implemented, therefore, our system provides significantly more possibilities to create fused images from various image sequences.

## 4.2 The Basic Solution

To ensure that the new development is understandable, this section summarizes some of the methods used in our basic system [6].

For feature detection point-like features, corner points were applied to ensure typical detection. The well-known Harris method was used [9]. Its Hessian matrix was determined by the components of the image's gradient at a given window, and its corner points were marked when both Eigen values of the matrix were higher than threshold.

Once the corner points have been identified in both images, the corresponding pairs in feature matching were determined by the following method. Normalized cross-correlation was calculated around the corner points, and where the value was the highest we could suppose the biggest similarity of the two matched points. The result of the pairing could have of course led to false results as well; therefore it was necessary to filter false pairs in our next step.

The transformation model estimation function is used to determine the transformation which creates the link between the pixels of the actual image and the pixels of the reference image. The transformation parameters were determined in the previous step by pairing the feature points, since the actual image and the reference image were taken with the same camera, so the homography transformation can be described as a matrix, which consists of eight free parameters (8 degrees of freedom) and it creates mappings between the points, where the points are given in homogeneous form and $c$ is the so called scaling factor and (12) [9].

$$
c \begin{bmatrix} x' \\ y' \\ 1 \end{bmatrix} = \begin{bmatrix} t_{11} & t_{12} & t_{13} \\ t_{21} & t_{22} & t_{23} \\ t_{31} & t_{32} & t_{33} \end{bmatrix} \begin{bmatrix} x \\ y \\ 1 \end{bmatrix} \tag{12}
$$

To determine the 8 parameters of the $T$ matrix, at least 4 points were needed, such that no three points were collinear. The incorrect pairings of the previous step were needed to be filtered, which was determined by using a RANSAC method [9]. Since the pictures were taken in different test conditions (indoors crowded environment and a much larger outdoor environment), so the adjustability of the RANSAC parameters was appropriate.

In the image resampling and transformation step the actual image was transformed by the $T$ matrix, and the resulting picture along with the reference picture was blended into one another by simple averaging.

## 4.3   Improved Version

The basic system used only one method for each step of the fusion. If the intensities fell into a narrower range at the input scene, and the Harris feature point detection failed to identify them sufficiently, then it was impossible to match the series of images. Therefore our developments are mainly concentrated on providing the users with more possibilities at the critical steps (Fig. 3). Therefore, we implemented several new approaches at the determination of feature points. Some of these methods do not only provide us with the coordinates of the point, but with the feature vector describing the environment. Thus, it was rational to use other techniques, such as the k-nearest neighbor method (KNN), which does not only explore the local intensity of the environment to locate pairs of feature points as the cross-correlation method, but it is able to use more characteristics. Finally, the desired quality was not always guaranteed by the original technique to blend images, since too harsh sub-boundaries were generated, so another option is provided with the improved system.

The user can choose from six types of feature point detector in the current system. The Moravec detector is based on the assumption [9] that a change in a

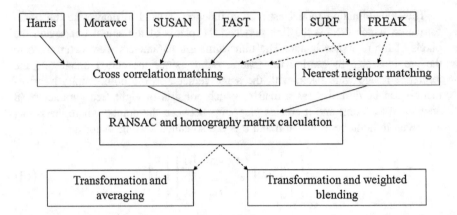

**Fig. 3** Workflow in the improved registration system

corner's intensity is great along all directions, while at an edge, it is only in one direction, and at a homogenous area it is negligible. Thus, the algorithm calculates the following quantities for each pixel: it determines the summation of squared differences (SSD) of the original local environment (window) and the in any direction translated window; and maps the smallest value to the examined point. Thus, the corner points will be the most intensified points as a result which can be found by searching for the maximum. The SUSAN [8] matches the intensity at the center of the mask with the intensity at any other points of the mask up to a certain threshold. The mask is generally circular, but occasionally it can be a rectangular. In the next step the number of similar intensities is determined. Finally, an evaluation expression determines the response of SUSAN operator.

FAST (Features from accelerated segment test) [10] is a corner point detector examines a circular part around the given point consist of 16 pixels, where the surrounding pixels are used in given order. If from a given number of the surrounding pixels are different from the given pixels up to a threshold than this pixel is a corner point.

SURF [9] detector is much faster and more efficient than its predecessor, while it has a much lower computational demand. It uses the integral of the images during the convolution step, simplifying them based on existing methods. It is invariant for rotation and scaling. To find the corner point, Hesse-determinant solution is used. A part of a local image is used to describe the environment of the feature points to calculate the Wavelet coefficients using Haar wavelets. Searching similar features it uses special indexes, which can significantly accelerate the search.

FREAK [11] "is also a binary descriptor and it evaluates 43 weighted Gaussians at locations around the key point. The pixels being averaged overlap, and are much more concentrated near the key point. This leads to a more accurate description of the key point. The actual FREAK algorithm uses a cascade for comparing these pairs, and puts the 64 most important bits in front to speed up the matching process" [11].

The SURF and the FREAK are not just feature detectors, since as a result they provide us with a feature vector (extractor), thus, while matching pairs, they do not only carry information about the location. This additional information can be processed by the K-nearest neighbor algorithm in order to match the image patches.

Instead of simply averaging the two inserted images, combining the image pixels with some sort of weighting can produce more realistic results. In our solution we determine the centroid of each image, and the distance between the given pixel and its centroid provides us with the magnitude of weighting. So we get a smoother transition while fitting.

## 4.4  Experimental Results

The presented system was implemented in .NET framework using the compute vision libraries AFORGE.NET, ACCORD.NET and OpenCV. The system is capable of loading thermal images and matching them to another one or to the result of an earlier fused image using more alternative methods and wide scale parameters. The user can examine continuously the partial results of the system, but the system is capable of fusing the reference and the actual thermal image in an automatic way. This is possible due to a prior set up of the system with similar images and the feature detector and its parameters were determined earlier.

The tests were run with sequences of thermal images sized 320 × 240 pixels. The images were taken at indoor and outdoor environments as well. In case of outdoor environment the applied RANSAC algorithm gave more stable results despite the non-deterministic nature of the method. Since the internal environment was mapped

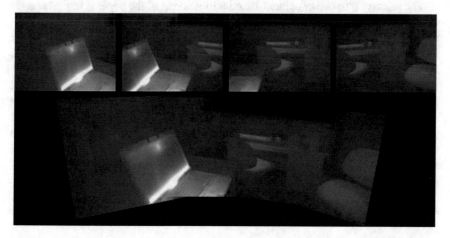

**Fig. 4** Thermal image series from local surroundings and the result of the registration: original images (*top*), results (*bottom*)

with larger step series and significant rotation was between the image details, in this case it had a greater impact on the alignment accuracy of the selected feature detection method and its setting. The different corner detectors provide results with different speed and different "sensitivity". If the local temperature distribution is more homogeneous than the traditional descriptors (for example the results of Moravec detector) gives better results combined with the RANSAC, as they give more responses than the SURF or FREAK method, and RANSAC can choose enough corresponding point pairs.

The software performs semi automatic thermal images fusion, so that in each step the user can choose from several alternatives. Since it presents a significantly larger space, it can be used effectively to examine a given workspace (Fig. 4).

## 5   Conclusions

The obtained results can be divided into three main groups, namely the enhancement and representation of patterns in thermal images, the methods of content-based retrieval of thermal images, and the procedure of thermal image fusion to promote the efficiency of object detection.

New canonical forms based on higher order singular value decomposition were produced to describe thermal images and highlight the pattern features. These canonical forms were applied for effective representation of thermal images, and enhancement of several specific features, as well as numerical reconstruction of thermal distribution changes, and compression of sequences. In addition, closely related to this study a method was developed to reconstruct the geometric properties of the examined surface.

The primary purpose of content-based retrieval part of the paper is to find images similar to a given image in a database containing thermal images. These algorithms use only the pure information from the image itself, neither textual indexes nor human contribution. The application possibility of Hu's invariants was analyzed, and the detection techniques of contour lines of object borders were examined.

The aim of thermal image fusion section is to make semi-automatic fusion of thermal images such that the user can choose from more alternatives in each step. A four-step algorithm fits the images: detection of feature points; pairing these points; production of transformation model; and merging of the original image and reference image in knowledge of the determined geometric transform. Since the result image represents a significantly larger space, so this method can be effectively applied for workspace analysis as well.

**Acknowledgements**  We acknowledge the financial support of this work by the Hungarian State and the European Union under the TÁMOP-4.2.1B-11/2/KMR-2011-0001 project.

# References

1. Rövid, A., Szeidl, L., Hashimoto, T.: Numerical reconstruction and compression of thermal image sequences. In: Fifth International Conference on Emerging Trends in Engineering and Technology (ICETET), pp. 298–302 (2012)
2. Szeidl, L., Várlaki, P.: HOSVD based canonical form for polytopic models of dynamic systems. J. Adv. Comput. Intell. Intell. Inf. **13**(1), 52–60. ISSN:1343-0130 (2009)
3. Szeidl, L., Baranyi, P., Petres, Z., Várlaki, P.: Numerical reconstruction of the HOSVD based canonical form of polytopic dynamic models. In: 3rd International Symposium on Computational Intelligence and Intelligent Informatics, Agadir, Morocco, pp. 111–116 (2007)
4. Sergyán, S.: Useful and effective feature descriptors in content-based image retrieval of thermal images. In: 4th IEEE International Symposium on Logistics and Industrial Informatics, LINDI, Smolenice, Slovakei, pp. 227–231 (2012)
5. Sergyán, S.: Edge detection techniques of thermal image. In: IEEE 10th Jubilee International Symposium on Intelligent Systems and Informatics, SISY 2012, Subotica, Serbia, pp. 55–58 (2012)
6. Vámossy, Z.: Thermal image fusion. In: IEEE 10th Jubilee International Symposium on Intelligent Systems and Informatics, SISY 2012, Subotica, Serbia, pp. 385–388 (2012)
7. Zitova, B., Flusser, J.: Image registration methods: a survey. Image Vis. Comput. **21**, 977–1000 (2003)
8. Szeliski, R.: Computer vision: algorithms and applications, p. 812. Springer, Berlin (2011)
9. Smith, S.M., Brady, J.M.: SUSAN—a new approach to low level image processing. Int. J. Comput. Vision **23**(1), 45–78 (1997)
10. Trajkovic, M., Hedley, M.: Fast corner detection. Image Vis. Comput. **16**(2), 75–87 (1998)
11. Alahi, A., Ortiz, R., Vandergheynst, P.: FREAK: fast retina keypoint. In: IEEE Conference on Computer Vision and Pattern Recognition (CVPR), pp. 510–517 (2012)

# The Doctrinal Base of Operational Employment of Air Defence Missile Units in the Light of the Relevant National and Allied Publications

Zoltan Krajnc

**Abstract** The study deals with the doctrinal base of operational employment of air defence missile units (subunits, combat groups) in the light of the relevant publications of the Hungarian national and allied forces. The author collected the determining doctrinal literatures from national and NATO documents. It was established that the integration of the air defence missile combat and its activities, as the basic activity of this branch of service into the air operations, has been doctrinally achieved. A commander of an air defence missile formation, its operational designer gets a logical, clear picture about the role and place of his troops from a doctrine.

**Keywords** Military operations · Military doctrine · Air defence · Air defence missile · Counter air operations

## 1 Introduction

The question of the proper effective operation of the military forces, their basic principles and the process of their actual application, the need for regulations of independent decision-making and creativity is almost as old as the appearance of the fighting forces. It was possible to notice, much longer before the beginning of the 20th century, that there was a need to examine the organization, build-up, application of the military by scientific means, but undoubtedly, that was the time when it became general practice.

In our days, in the documents and documentation systems that are called doctrines within the alliance, basically a change may occur concerning the entire structure, weaponry, or in case of change in the political or military circumstances. The same happened after the 1991 Gulf war when in the USA Air Force, on doctrinal level, the old was "swept out" and the new AFDD system (Air Force Doctrine Documents) was introduced.

Z. Krajnc (✉)
The National University of Public Service, Budapest, Hungary
e-mail: krajnc.zoltan@uni-nke.hu

© Springer International Publishing Switzerland 2016                    177
L. Nádai and J. Padányi (eds.), *Critical Infrastructure Protection Research*,
Topics in Intelligent Engineering and Informatics 12,
DOI 10.1007/978-3-319-28091-2_15

When we joined NATO, the Hungarian Air Defence Missile and Artillery was in a similar, but at the same time, completely different situation, because basically, the military technology remained the same.[1] Since the more modern Mistral system integrated, however, a foundation for the doctrine needed to be established. They were necessary for application independently, with all arms included, nationally, with NATO troops, and for its preparation.[2]

The process was coloured by the fact that the air defence artillery was integrated into one military unit (brigade, later regiment), which, from certain point of view, could simplify the introduction of doctrinal regulations.

After the change of regime, the notion of the doctrine, its development has become the central issue of the research in the military science, since the capability of interoperability of our forces, the elements they consist of, is based on intellectual compatibility. The main element of this compatibility, the main tactical and strategic application, their main theories are laid out in the so-called allied "operational doctrines". It means if we want to meet the requirements of our allies, we have to put into force the allied doctrines, and we need to take appropriate action accordingly. We also need to create a so-called national doctrinal literature system where we record military operations based on national interests and their main theoretical description.

Before the examination of the question, which constitutes the main part of our article, it would be necessary to clarify the notion and the pivotal issues of a doctrine.

A doctrine, according to the general formulation, is a system of a study subject, scientific theories and different views and according to its contents and direction, it can be a political, ideological, theological, economic, financial and military doctrine. Military doctrines are worked out, accepted, applied system of documents, based on which the aims established in the national security and in national military strategy could be realized and the tasks could be performed.

The essence of a military doctrine can be interpreted in a narrower and a wider sense[3]: in the wider approach there is a summary of theoretical and practical knowledge that needs to be used for the successful application of a military force, so it needs to be viewed as a kind of a knowledge system. From the narrower point

---

[1]It remained unchanged, if it can be said that anything remained, since the previous air defence troops, after "vegetation" that took only a couple of years, became completely disbanded and according to the old terminology, the "troop air defence missile and artillery troops" branch remained in the system, although only partially (Author).

[2]The special area, similarly to the other components, remained unregulated, since the old "Warsaw Contract Armed Forces" type regulations became irrelevant or they could not be synchronized with the NATO doctrines.

[3]A military doctrine is the official term of a way of thinking which is accepted by the military as relevant in a given period of time, and it can be applied in the existing and future conflicts, for their preparation and for application to achieve success. WILLCOKS, M.A.: Future conflict and military doctrine, In.: RUSI JOURNAL 1994/3., 7 pages.

**Fig. 1** Doctrinal determinants of operational employment of air defence missile units

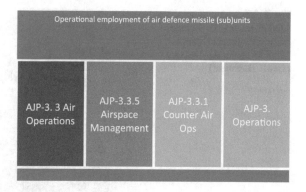

of view, it is a kind of a book of regulations, including rules, basic guides and guidelines for the entire military force or for individual segments (Fig. 1).

To simplify, there is a classic whole-part relation between the wider and narrower interpretation. The "whole" is viewed as the entire doctrine structure, the entire doctrinal literature, by its global study we can understand the generally accepted, relevant system of military science. Some doctrines, as part of the entire, describe one component of the operational application, their quantitative summary gives the wholeness, in other words, the knowledge of warfare.

In our article we want to tackle the application, and the doctrinal preparation of air defence artillery subunits (units, combat troops). We have summarised those national and allied regulations that may assist, and at the same time define, our combat operations and the process of preparation.

## 2 The Place and Role of Air Combat and Operational Activities of Air Defence Missile Units Within the Operations

It can be established, even after a superficial examination, that includes theoretical basis of small elements, that there is no relevant difference between the examined national[4] and NATO doctrines.[5] Consequently, their examination is based on not the traditional "national—allied" dichotomy, but is built on the specifications of air

---

[4]Hungarian doctrines: Doctrine for Joint Operations, Operational Doctrine of Hungarian Home Defence Forces (draft), Air Force Doctrine, Counter Air Operations (Magyar Honvédség Összhaderőnemi Doktrína; Magyar Honvédség Összhaderőnemi Hadműveleti Doktrína (tervezet); Légierő Doktrína; Légi Szembenállási Doktrína).

[5]AJP-3.3 Air Operations; AJP-3.3.1 Counter Air Operations; AJP-3.3.5 Airspace Control, etc.

**Fig. 2** Weapon systems of
12nd Arrabona Air Defence
Missile Regiment (Source:
12nd Arrabona Air Defence
Missile Regiment)

defence artillery missile, on a more interesting content of "place—role—operations
—leadership and command—guidance—coordination".

From the point of view of air defence artillery troops, the starting basis of the
elements of combat activities of the air force, is the category of "air defence
combat" and it can be found not only in national doctrines, and as we see, it does
not present any "disharmony" with the allied categories. It fits into operations and
into operational forms within the system of NATO, accepted by consensus (Fig. 2).

"Air defence combat means activities that destroy or limit enemy air defence attack devices by combat against specific districts, objects as well as protection of certain combat groupings".[6]

The doctrine writers[7] point out that the air combat, in all cases, can be considered an active reaction to the enemy air attack and in spite of the fact that its execution is an attack with the aim of defence, "it shows activities including manoeuvres and fire delivered by the applied air defence missiles and artillery fire weapons, weapon systems."[8]

The air defence combat as a "core particle" fits into the categories of counter air operations and defensive counter air operations. Since counter air operations and defensive counter air operations are executed by the same forces, in the same air space, so in most cases it is impossible to separate them. The present situation, as well as the commander's decision, would determine, in all cases, whether in the activities of the friendly troops the attack or the function of defence should be the dominant one.

As a definition, the notion of the counter air can be described as "operations in one dimension of warfare aimed to destroy the enemy air force that can be interpreted as a part of the operations in the air". Defensive counter air, as a collective notion, includes those regulations, processes and devices whose effect on the enemy's static and rotary air devices as well as on their UAVs would diminish and destroy the success of their intelligence, attack and defence capabilities. The execution of counter air operations—which include mainly the activities of fighters, fighter-bombers, air defence missiles and artillery weapons as well as the electronic counter activities—can provide the necessary maintenance level of control over the air space and it is possible to establish favourable conditions for further operations of friendly troops.

Doctrines clarify that defensive counter air is a synonym for air defence. It can be considered as an operational system which means that it is a reaction to the air attack of the enemy. It integrates those devices and activities aiming to limit the effectiveness of enemy air attacks. Accordingly, the notion of air defence combat belongs into the category of air opposition.

---

[6]Air Force Doctrine of Hungarian Home Defence Forces (MH DSZOFT kód: 13013) Nyt. szám: 563/614/2004./LEP (in page 39).

[7]The author, as specialist, working at the Joint Operational Department of Military Sciences and Officer Training Faculty of National University of Public Service, took part in the writing of doctrines and in the adaptation of the relevant NATO publications.

[8]Air Force Doctrine of Home Defence Forces (Hungary) (MH DSZOFT kód: 13013) Nyt. szám: 563/614/2004./LEP (in page 41).

## 3   The Notion of Air Defence, the Combat Role and Place of Air Defence Missile Units in the System of Air Defence

To establish the notion of air defence in doctrines is an important part, whose aim is to defend certain areas, objects from the air, as well as to repel enemy attacks at all times, to provide means for the effective activities of friendly troops and to safeguard our own values and forces.

It is obvious that the new attack devices need the extension of the traditional notion of the air defence, so in doctrines and in the system of tasks the notion of extended air defence has appeared. It integrated into one system those activities, regulations and devices that apply to aircraft, helicopters, UAVs and remote controlled air defence attack devices, as well as actions against combat aerodynamic and ballistic missiles.

The all-arms view of the combat means that that the application of all arms is harmonized and is based on common concept. Consequently, since reaching the aims of air defence is possible not only by the available sources for the air defence, but they can be supported by land forces and the navy, and requires the application of all available weapon systems. As a result of this integration, among other benefits, the air defence can be realized on a higher level, the so-called "extended, integrated air defence".

The components of the air defence are activities executed within the framework of air defence and they can be categorized into four groups: Active air defence, passive air defence, previous operations and combat group command operations. From the point of view of the air defence missiles, active air defence is the most important category, within its framework with the support of the modern and secure command and control systems the tasks of the weapon and sensor systems is to recce enemy aircraft and missiles while in flight, to intercept and destroy them.

When we examine the operational area of the active air defence, it is necessary to emphasize the importance of the airspace checking, the activities of the air space checking troops, since the speed of aircraft and their nationality identification, as well as the security of data forwarding can considerably increase the survival of friendly aircraft, as well as the effectiveness of defence against enemy air attack devices.

The establishment of tasks for subunits or units allocated to the defensive air opposition, especially the active defence, the indication of objects and areas to defend can happen only in centralized way. Depending on the combat task, on the size of the defendable object or area 4 types of air defence can be distinguished: the district or area protection, point protection, self-defence, and defence of air objects.

From the point of view of air defence missile troops the categories of district or area protection, point protection, and self-defence need to be given.

During district or area protection the air defence tasks are executed in an area outlined by geographical coordinates, according to the integrated air defence, by using different weapon systems where the main aim is to keep the integrity of a given district against an enemy air attack.

Point protection means protection of a relatively small geographical area. It can be executed as protection for smaller objects, combat groupings or for especially important combat elements.

Self-defence as the basic function provides further the maintenance of the further combat capability. In this case the task is to protect friendly troops against the direct enemy attacks, with the help of weapon systems, their combat coordination is the task of a unit's or subunit's commander.

# 4 Summary

In this study I made a quick overview of doctrine-based application of the air defence missile subunits (units) and as a summary, we can say that the "doctrinal support" of this special area is organized.

The integration of the air defence missile combat and its activities, as the basic activity of this branch of service into the air operations, has been doctrinally achieved. A commander of an air defence missile formation, its operational designer gets a logical, clear picture about the role and place of his troops from a doctrine.

A smaller problem is that on the level of weapon systems there are no doctrinal handbooks, "small" doctrinal regulations. A principal regulation, like FM 44-100 can be found only in the inner regulations of a regiment (for example, SOPs[9]) However, it does not pose any difficulties in the training of the regiment's personnel, since the regiment, as an air defence missile unit, is unique in the HDF. Consequently, the need for a very special doctrine applies for only one unit. It is obvious, that specialists and organisations who deal with operational issues of combat forces would enlarge their intellectual horizon, assist in operations planning, if a doctrine of this kind existed.

I would like to add another thought to this issue. The fact that practically it was the personnel of the regiment who established the majority of theoretical principles themselves, with the help of the synchronized application of air defence missile artillery subunits and battalions. The adaptation of NATO doctrines, for example, FM 44-100, as well as seeing foreign examples, was of considerable assistance. This fact is a serious weapon fact, since the establishment of doctrines is not a task for any member state, but their task is to carry them out, taking them into consideration during combat and training.

**Acknowledgement** We acknowledge the financial support of this work by the Hungarian State and the European Union under the TÁMOP-4.2.1B-11/2/KMR-2011-0001 project.

---

[9]SOP—Standing Operating Procedures.

# References

1. AAP 6 NATO Glossary of Terms and Definitions/Glossaire OTAN de termes et définition. http://www.nato.int/docu/stanag/aap006/aap6.htm
2. AJP-3 (A) Allied Doctrine For Joint Operations, July 2007
3. AJP-3.3(A) Allied Joint Doctrine For Air And Space Operations, November 2009
4. RUTTAI LÁSZLÓ – KRAJNC ZOLTÁN: A légierő doktrinális alapjai, Budapest, ZMNE, egyetemi jegyzet (2001)
5. WILLCOKS, M.A.: A jövő konfliktusa és a katonai doktrína, In: Rusi Journal (1994/3)
6. KRAJNC ZOLTÁN: A magyar légierő doktrínáját befolyásoló tényezők komplex vizsgálata, PhD-értekezés, Budapest, ZMNE Egyetemi könyvtár (2000)
7. Magyar Honvédség Légierő Doktrínája (MH DSZOFT kód: 13013) Nyt. szám: 563/614/2004./LEP
8. A légi szembenállás doktrínája (MH DSZOFT kód: 13021) Nyt. szám: 563/641/2004./LEP
9. Harcászati Légvédelmi Doktrína (215/2006. (HK 23.))
10. Varga László: A légvédelmi rakéta erők alkalmazásának időszerű kérdései. http://www.szrfk.hu/rtk/kulonszamok/2008_cikkek/Varga_Laszlo.pdf

Printed in the United States
By Bookmasters